W9-BIC-638

Growing Up with SCIENCE®

Third Edition

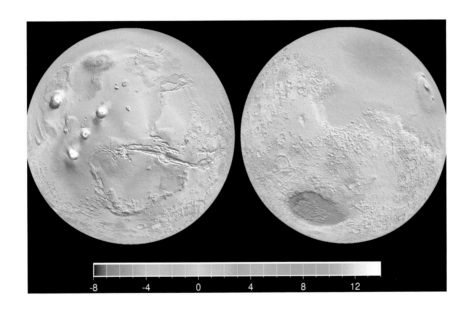

-8   -4   0   4   8   12

8

Mapmaking–Mining and quarrying

Marshall Cavendish
Reference
New York

Marshall Cavendish
99 White Plains Road
Tarrytown, NY 10591

www.marshallcavendish.us

© 2006 Marshall Cavendish Corporation
© 1987, 1990 Marshall Cavendish Limited

GROWING UP WITH SCIENCE is a registered trademark
of Marshall Cavendish Corporation

Library of Congress Cataloging-in-Publication Data

Growing up with science.— 3rd ed.
    p. cm.
  Includes index.
  Contents: v. 1. Abrasive-Astronomy — v. 2. Atmosphere-Cable television —
v. 3. Cable travel-Cotton — v. 4. Crane-Electricity — v. 5 Electric motor-
Friction — v. 6. Fuel cell-Immune system — v. 7. Induction-Magnetism —
v. 8. Mapmaking-Mining and quarrying — v. 9. Missile and torpedo-Oil
exploration and refining — v. 10. Optics-Plant kingdom — v. 11. Plasma
physics-Radiotherapy — v. 12. Railroad system-Seismology — v. 13.
Semiconductor-Sports — v. 14. Spring-Thermography — v. 15. Thermometer-
Virus, biological — v. 16. Virus, computer-Zoology — v. 17. Index.
  ISBN 0-7614-7505-2 (set)
  ISBN 0-7614-7513-3 (vol. 8)
  1. Science—Encyclopedias.

Q121.G764 2006
503—dc22

                                        2004049962
                          09 08 07 06 05 6 5 4 3 2 1
Printed in China

CONSULTANT

Donald R. Franceschetti, Ph.D.

Dunavant Professor at the University of Memphis

Donald R. Franceschetti is a member of the American
Chemical Society, the American Physical Society, the
Cognitive Science Society, the History of Science Society,
and the Society for Neuroscience.

CONTRIBUTORS TO VOLUME 8

Chris Cooper   Freddy Tipple

John Farndon   Emma Young

Tom Jackson

Marshall Cavendish

Editor: Peter Mavrikis

Editorial Director: Paul Bernabeo

Production Manager: Alan Tsai

The Brown Reference Group

Editors: Leon Gray and Simon Hall

Designer: Sarah Williams

Picture Researcher: Helen Simm

Indexer: Kay Ollerenshaw

Illustrators: Darren Awuah and Mark Walker

Managing Editor: Bridget Giles

Art Director: Dave Goodman

CONTENTS

KEY TO COLOR CODING OF ARTICLES

EARTH, SPACE, AND ENVIRONMENTAL SCIENCES

LIFE SCIENCES AND MEDICINE

MATHEMATICS

PHYSICS AND CHEMISTRY

TECHNOLOGY

PEOPLE

Mapmaking

Maps are drawings of the world, or parts of it. They are greatly reduced to fit on a sheet of paper. Maps contain a lot of information, in the form of symbols, words, lines, and colors. One map on a sheet of paper may contain as many facts about a place as a book.

The many different kinds of maps include atlas maps, showing continents and countries, and general reference (topographic) maps. Topographic maps show natural and human-made features, such as lakes and rivers, hills and mountains, roads and railroads, and cities and villages. The United States Geological Survey (U.S.G.S.) is responsible for plotting most maps in the United States.

There are many kinds of special maps. Some, such as road maps, are made for motorists. Maps for outdoor pursuits feature information about amenities, campsites, and trails. Geological maps show the different kinds of rocks and rock structures. Weather maps are used by forecasters to study the weather. Navigational charts are maps of coastal waters, showing underwater features.

▼ *This map of the Americas has been reproduced from a seventeenth-century atlas. Surveying and mapmaking were not very accurate at that time, so this map looks very different from modern maps.*

Drawing to scale

Maps must be drawn to scale or they would be very confusing. This means that a given distance on a map must correspond to a specific distance on the ground. For example, if a topographic map is drawn to a scale of 1:63,360, it means that 1 inch (2.54 centimeters) on the map is equal to 63,360 inches (160,934 centimeters), or 1 mile (1.6 kilometers), on the ground. If a person using the map wants to find the distance between two points on the ground, they first need to measure the distance between the points on the map. For example, if this distance is 5 inches (12.7 centimeters), the distance across the ground ("as the crow flies," in a straight, horizontal line) is 5 miles (8 kilometers).

There are three ways that mapmakers show the scale of reduction on a particular map. Some maps have only one legend, and other maps will use all three types. A scale such as 1:63,360 is called a representative fraction. A graphic scale is a bar divided into distances representing numbers of miles or kilometers on the ground. The third scale is more simple: the legend will state something such as "1 inch = 50 miles."

Mapmakers use both small-scale and large-scale maps. Small-scale maps, seen in atlases, cover large areas, such as countries and continents. However, they leave out many details because there is no space to show them. For example, a world map may have a scale of 1:100,000,000. On this map, 1 inch (2.54 centimeters) is equal to 1,578⅓ miles (2,540 kilometers) on the ground. Large-scale maps cover much smaller areas and show far more details. The largest-scale maps are called plans. A plan may show only one street, including each house, the land associated with each house, and the boundaries between houses.

Latitude and longitude

Most maps are sectioned off by networks of curved or straight lines. These lines help map users locate places. Such networks may form a simple grid of equally spaced vertical and horizontal lines. On many maps, however, the networks, called graticules, consist of lines of latitude and longitude.

▲ *A surveyor uses a theodolite to measure spot heights and determine coordinates of an area of land being mapped. Creating maps by surveying in this way is a long and laborious process.*

The position of every place on Earth can be defined using latitude and longitude. Together, these measurements are called coordinates. Lines of latitude (parallels) are imaginary lines running from east to west around the world. They are parallel to the equator—an imaginary line that circles Earth exactly halfway between the North and South poles. Lines of latitude are measured in degrees, minutes (60 minutes equals one degree), and seconds (60 seconds equals one minute). The equator is 0 degrees latitude, the North Pole is 90 degrees north, and the South Pole is 90 degrees south. The latitude of any point on Earth can be

Lines of longitude are all the same length. However, lines of latitude become shorter toward the poles. The total distance around Earth at the equator is 24,902 miles (40,075 kilometers). At the equator, one degree of longitude is $69^{17}/_{100}$ miles (113.32 kilometers) long, one minute of longitude is $1^{3}/_{20}$ miles (1.85 kilometers), and one second is 101 feet (30.8 meters).

Map symbols and distortions

Within the networks, maps show the shapes and features of the land. To fit as much information as possible on a map, cartographers (mapmakers) use symbols for many features. For example, an area of trees may be depicted by evenly spaced green symbols that resemble fir trees. Colors are important in making maps clear. Human-made features, such as roads, are often shown in red and black, water features in blue, plants in green, and the relief (land height) in brown.

The main way of showing relief is by contours— imaginary lines joining places of the same height. Together with contours, the heights of hilltops or mountain peaks are given in figures. These are called spot heights. Other maps use various kinds of shading to depict hills and valleys. To produce some atlas maps, the mapmakers make a large plastic model of the area, showing all the mountains and valleys. The scale of heights on these models is exaggerated to make the features stand out more. The model is then painted and lit up so that shadows are created to further emphasize the ups and downs of the land. The model is then photographed and the photograph is used as a base for the map, onto which other features and names can be added.

Some features that are important to map users must be exaggerated and drawn to a false scale, because otherwise they would be invisible. For example, the finest line that appears on maps is about $1/_{500}$ inch (0.005 centimeter) wide. Using the

measured as an imaginary angle formed by drawing one line between the center of Earth and the equator and another line from Earth's center to that point. For example, the latitude of Kansas City, Kansas, is 39 degrees north.

Lines of longitude (meridians) encircle Earth, running at right angles to lines of latitude. Lines of longitude run from north to south and pass through the poles. The angular distance around Earth is 360 degrees—the number of degrees in a circle. However, lines of longitude are measured in two half-circles: 180 degrees west and 180 degrees east of the prime meridian (0 degrees longitude). By an international agreement made in 1884, the prime meridian runs through Greenwich in London, England. The longitude of Kansas City is 94 degrees, 40 minutes west. This means that it lies at an angle of 94 degrees 40 minutes west of the prime meridian, as measured from the center of Earth in a clockwise direction.

small scale of 1:1,000,000, this line represents a distance of 166⅔ feet (50.8 meters) on the ground. This means that important rivers and roads would be represented by lines that could be seen only through a microscope if they were drawn accurately to scale.

Mapmaking

The first stage in mapmaking on Earth is surveying. Surveyors work on the ground, measuring distances, angles between points, directions, and heights. They begin the mapping of an uncharted region by deciding on a network of points arranged in triangles. They then measure the distances and angles between the points. To fix the latitude and longitude of the points, surveyors have historically observed the stars. With the advance of satellite technology, however, it is now common for surveyors to use a portable Global Positioning System (GPS). By triangulating signals from satellites, these devices can quickly pinpoint a location to within 3 feet (1 meter).

When a very accurate framework has been completed, secondary points are fixed and the details of the land are filled in. After World War II (1939–1945), detailed mapping largely relied on

photographs taken from the air. The science of making measurements from photographs is called photogrammetry. All these measurements are based on points upon which coordinates (longitude and latitude) and heights have been fixed by surveyors on the ground. These points can be identified on the photographs.

Aerial photographs are taken in strips. Each photograph overlaps the one before by about 60 percent. These overlaps can be looked at through stereoscopes, which allow the viewer to see a three-dimensional picture of the land. When the photographs are correctly positioned, all the details can be plotted. Because of the three-dimensional effect, contours can also be drawn.

◄ *Remote areas, such as this swampland in Florida, are difficult to map using traditional methods of surveying. It was not until the use of aerial photography in the early twentieth century that these areas could be mapped with any degree of accuracy. Even today, many parts of the world are not mapped.*

More recently, space photography from satellites has become widely used. In particular, improved weather maps have been made possible by photographs of cloud cover taken from space.

The advent of the computer has meant that many maps can now be plotted on computer, rather than by hand. With electronic surveying and sensing techniques (such as laser range-finders and radar surveying) and digital satellite photography, information can be transferred directly to a computer where it is stored, correlated, and mapped using specialized software. Once the information is entered into the computer, it can quickly be used to create different types of maps.

Map projections

Earth is a spheroid (not quite round) that bulges slightly at the equator and is flattened at the poles. Although Earth's surface is curved, it is possible to map a small area without distortion. However, an allowance for the curvature must be made when mapping large areas. For example, if a sheet of tracing paper is placed on a globe—the only true representation of the entire Earth—it is easy to

trace details of a small area. However, a large continent cannot be traced without creasing and crumpling the paper. To solve this problem, mapmakers have devised various ways of using the latitude and longitude lines to produce flat drawings of Earth's curved surface. These drawings are called projections.

Some simple map projections are called perspective projections. Imagine a glass sphere, representing Earth, on which a graticule has been engraved. With a light at the center of the sphere, shadows of the grid would be cast onto any flat surface, such as a sheet of paper.

The cylindrical projection is developed as though a paper cylinder had been wrapped around the sphere, touching it along the equator. The other lines of latitude will be cast as shadows onto the cylinder, but the distances between them will increase as they get farther away from the equator. Lines of longitude will appear as straight lines. These lines will be parallel and will not meet up, as they should, at the North and South Poles. Therefore the cylindrical projection is accurate only at the equator. Away from the equator, it becomes increasingly distorted.

A second kind of perspective projection is the azimuthal projection. In this case, a flat piece of paper is placed on the globe, touching it at only one point, usually one of the poles. In azimuthal projections, the distortion increases away from the point of contact.

The third type of projection, the conical projection, is based on the idea of a paper cone being placed over the glass sphere. If the tip of the cone is directly over the North Pole, the sides of the cone will touch the globe along one of the lines of latitude in the Northern Hemisphere. The grid cast

◀ *This topographical image of Honolulu, on the island of Oahu, Hawaii, was created using a Landsat satellite image combined with Shuttle Radar Topography Mission (SRTM) topographical information. SRTM consisted of a specially modified radar system that flew onboard the Space Shuttle* Endeavour *during an 11-day mission in February 2000. SRTM obtained elevation data on a near-global scale to generate the most complete high-resolution digital topographic database of Earth.*

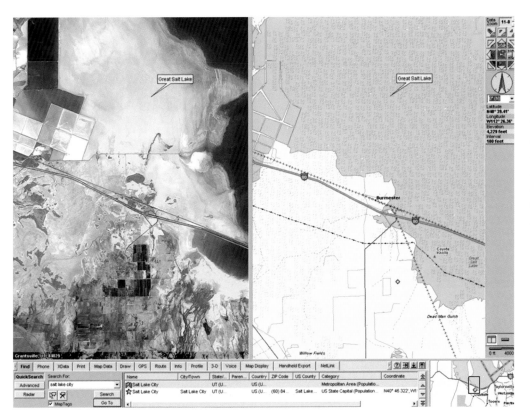

◀ *A cartographer creates a map using DeLorme XMap mapping software. By using an updatable database of different information gathered from surveys, the software helps cartographers create a number of different types of maps of an area more easily than by using conventional methods.*

onto the paper cone will be true only along the line of latitude where the paper touches the sphere. Distortion increases to the north and south.

Some globes are covered by strips of paper called gores. These strips are joined at the equator but taper toward the poles. If they are removed, they form a type of "interrupted" map projection. However, such projections are of little use, because continents and oceans are split into sections.

Since they are distorted, pure perspective projections are seldom used in mapmaking. However, many map projections are forms of perspective projections that have been modified mathematically to reduce the distortion. Some projections are developed entirely by mathematics and owe nothing to perspective projections. These are called conventional projections and are often used for world maps.

In choosing a projection, cartographers must decide what properties they want to preserve on the map. For example, maps can preserve correct areas, shapes, angular relationships and directions, and distances. However, no projection can preserve all of these features within one map of a large area.

Modern printing techniques

Until the 1960s, maps were mainly drawn on paper with black ink, or they were directly engraved on copper plates, which were used for printing the maps. Later, many high-quality maps were drawn on a coated plastic film using a pointed stylus. This technique, called photoengraving, gives a much higher quality than penwork and is easier than engraving. When light is shone through the film, a negative image is produced from which a positive can be made photographically. Symbols and names are printed on separate strips of film and glued in position on the positive. For colored maps, each color is prepared separately as an overlay and separate plates are made.

With many maps now created using computer systems, they can be copied and distributed as digital files that can simply be displayed on a screen or printed directly from the computer onto paper. Maps produced using computer technology are called orthophotomaps.

See also: GEOGRAPHY • SURVEYING

Marconi, Guglielmo

Italian inventor Guglielmo Marconi developed wireless telegraphy and later radio communication and broadcasting. Radio waves had already been discovered by German physicist Heinrich Hertz, but it was Marconi who developed Hertz's ideas and became the first person to send a radio signal outside the laboratory.

Guglielmo Marconi was born in Bologna, Italy, on April 25, 1874. Throughout most of his childhood, Marconi was educated by private tutors, although he did receive formal instruction in physics at the Technical Institute in Livorno for a brief period. There, he studied under many distinguished Italian professors and took a keen interest in the work of prominent physicists such as Scottish physicist James Clerk Maxwell (1831–1879), German physicist Heinrich Rudolph Hertz (1857–1894), and English physicist Oliver Joseph Lodge (1851–1940), among others. However, Marconi was very mechanically minded and preferred tinkering with machinery to his studies. Much to his father's annoyance, he decided not to attend university.

In 1895, Marconi moved to his father's country estate at Pontecchio. There, he set up a laboratory dedicated to the study of radio waves, which had been discovered by Hertz in the 1880s. Marconi filled the attic rooms of his laboratory with equipment such as a coil and spark to transmit the radio waves and a receiver to pick them up. Soon he was able to demonstrate how, if he pressed a key at one end of the laboratory, a buzzer would sound 30 feet (9 meters) away.

▶ *Marconi sent the first radio signals across the Atlantic Ocean using this experimental tuned transmitter. Later, Marconi established a worldwide radio telegraph network for the British government.*

Marconi gradually developed and improved his equipment so that he could send signals farther and farther, eventually transmitting and receiving radio signals up to a distance of 1½ miles (2.4 kilometers). He even discovered that it did not matter if there was a hill between the transmitter and receiver.

Move to England

No one in Italy would fund Marconi's experiments, so in 1896 Marconi took his apparatus to England. With help from William Preece (1834–1913), engineer-in-chief of the British Post Office, Marconi was soon sending signals over a distance of 9 miles (14 kilometers). Later that year, Marconi was granted a patent for his system of wireless telegraphy. In 1897, following a series of successful demonstrations in London, on Salisbury Plain, and across the Bristol Channel, Marconi formed The Wireless Telegraph & Signal Company Limited, which was renamed Marconi Wireless Telegraph Company Limited in 1900.

▶ *This portrait of Guglielmo Marconi was taken at Signal Hill in Newfoundland, Canada, in 1901. He is surrounded by the instruments with which he received the first wireless signal across the Atlantic Ocean from Poldhu in Cornwall.*

There followed a number of successful demonstrations of wireless telegraphy, most notably in December 1901, when he made the first wireless communication across the Atlantic Ocean—from Poldhu in Cornwall, England, to St. Johns in Newfoundland, Canada. By doing so, Marconi proved that wireless communications were not affected by the curvature of Earth.

A commercial success

As wireless telegraphy began to be developed commercially, Marconi was very much involved in the process. He patented many improvements to his system and several new inventions, such as the magnetic detector (1902), which became the standard wireless receiver for many years, and the horizontal directional aerial (1905). Following a number of smaller-scale enterprises, Marconi went on to establish the first transatlantic commercial service between Glace Bay in Nova Scotia, Canada, and Clifden, Ireland, in 1907. Thereafter, it was installed in many different places—including ships such as the *Montrose,* where a ship-to-shore message led to the arrest of a murderer in 1910.

Later life

During World War I (1914–1918), Marconi joined the Italian army but later transferred to the navy. As a naval commander, Marconi experimented with short-wave communications. After the war, Marconi set up a series of trials, which culminated in the beam system for communications over long distances. The first beam station, linking England to Canada, opened in 1926.

In the 1930s, Marconi worked on wireless communications using microwaves, establishing the first microwave link between the Vatican City and the Pope's summer retreat at Castel Gandolfo. His work in this field, which went on to form the basis of radar, was cut short when he died in Rome, aged 63. His death was marked by a two-minute radio silence worldwide.

Marconi received many honors and awards during his lifetime, most notably the 1909 Nobel Prize for physics, which he shared with German physicist Karl Braun (1850–1918).

See also: RADIO • TELECOMMUNICATIONS

Marine propulsion

Marine vessels move through the water using muscle power, the wind, or engines. Large modern ships usually have propellers powered by diesel engines or gas turbines. There are many different shapes and sizes of propellers, depending on the needs of the different types of boats and ships.

The earliest boats were propelled (moved) by paddles and oars, using human muscle power. An oar is simply a lever. It rests in a notch on the side of the boat, and this acts as a fixed support. When a forward-moving pressure is put on the handle of the oar, the blade pushes back against the water, forcing the boat forward. The paddle used to propel a canoe works the same way, except that the canoeist's own arm acts as a moving support.

Sails

A sailing boat relies on the wind to push it through the water. The earliest sails were made of animal skins or flattened, woven reeds. Later, sails were made of strong woven cotton called canvas. Modern sails are usually made from strong synthetic fabrics, such as polyester. These synthetic fibers do not absorb much water and will not rot.

The sails of the boat are hung from one or more masts by a system of ropes or wires known as the rigging. There are two main ways in which sails are usually arranged. A square-rigged boat has sails set facing the direction of movement. Fore-and-aft rigging uses sails set along the length of the boat.

In a square-rigged boat, the wind simply pushes the boat along by blowing against the back of the sails. A fore-and-aft sail works in a more complex way, allowing the boat to be sailed "close to the wind"—pointing almost straight into the face of the wind. When the wind strikes the sail, the sail curves out and resolves (divides) the power of the wind into several forces. One force drives the boat forward. Another force pushes at right angles to the boat and tends to make it slip sideways and tip over. This force is balanced by the resistance of the ship's hull (body) and keel (lengthwise supporting timber). The crew of a small boat have to balance the tipping force by sitting on the side of the boat toward the wind.

▶ The steam ship **Washington,** built in the 1840s, was one of the first large steam vessels. The ship's engine had two 72-inch (183-centimeter) steam cylinders and was used to power large paddle wheels. These paddle wheels were supplemented by sails. **Washington** was used to carry passengers across the Atlantic from the United States to Europe.

Engines

An engine is used to drive a propulsion device, such as a paddle wheel or propeller. The first engine-powered boats used steam engines. During the late eighteenth century, many experimental boats were tried out in Britain, France, and the United States.

French physicist and inventor Denis Papin (1647–c. 1712) had drawn up plans for a boat with a paddle wheel that was driven by a simple steam engine in 1690. It was not until 1783, however, that the first paddle steamer was built, by two French noblemen, the Marquis de Jouffroy d'Abbans and the Comte Charles de Follenay. The boat, named *Pyroscaphe*, sailed up the Saône River in France for 15 minutes before its steam engine blew.

Soon after, in 1786, U.S. inventor John Fitch (1743–1798) successfully tested a steam-powered boat on the Delaware River. Fitch's vessel, which had six paddles on each side, traveled at a speed of 3 miles (5 kilometers) per hour.

▲ *This large diesel engine is used to power the propellers of a cargo ship. Diesel engines in ships can vary in size from small engines, similar to those found in automobiles, to engines such as this one, which must be housed in huge engine rooms.*

Around the same time that Fitch was working in the United States, William Symington (1763–1831) was working in Scotland. Symington's first steamer sailed on Dalswinton Lake near Dumfries in 1788. In 1801, Symington built the *Charlotte Dundas*, which was named for the daughter of Symington's financial backer. This boat had a single paddle wheel in the stern and was powered by a double-action engine—one in which both the up and down strokes of the piston do the work. This type of engine had been invented by fellow Scot James Watt (1736–1819). The *Charlotte Dundas* was the first practical steamboat, because in 1803 it towed two loaded vessels for 19½ miles (31.5 kilometers) along the Forth and Clyde Canal in Scotland.

One of the people interested in the *Charlotte Dundas*, who studied the boat carefully on a visit to Scotland, was U.S. engineer Robert Fulton (1765–1815). After his return to the United States, Fulton built a steamship he called the *Clermont*. It had side paddles and was powered by a Boulton and Watt engine from England. The 150-foot (480-meter) boat went into service in 1803 to carry paying passengers on the Hudson River in New York. On its first trip, it went 150 miles (240 kilometers) from New York City to Albany, taking 32 hours going up and 30 hours coming back. The *Clermont* was a great success, and steamship travel began to grow in popularity on all the chief rivers of the United States.

The first crossing of the Atlantic by a real steamship took place in 1838. Two English steamship companies were racing to have the first ship to reach New York from England. The *Sirius* arrived first, having taken 18½ days for the crossing. However, the *Great Western* was really the faster ship. Although it arrived in New York a few hours after the *Sirius*, the *Sirius* had in fact left England four days earlier.

In the late 1800s, steam engines began to give way to steam turbines. The first steam-turbine ship was the yacht *Turbinia*, which was built in 1894. Most

DID YOU KNOW?

The first propeller-driven boat was built in 1894. It was called the *Archimedes* for the scientist named Archimedes (287–212 BCE). His work had a lasting effect on marine propulsion. He is credited with the invention of the screw, often called the Archimedean screw. His screw pump, created to pump out flooded ships and for supplying water for agricultural purposes, was the forerunner of the screw propeller.

modern ships are powered by steam turbines or diesel engines. Small boats often have gasoline engines, and some high-speed boats are driven by gas turbines. Non-nuclear submarines are powered by electric motors underwater. On the surface or at periscope depth, they are powered by electric generators driven by diesel engines. The generators also recharge the batteries. A nuclear submarine uses heat from a nuclear reactor to make steam, which drives the submarine's steam turbines.

The early steamboats were moved by large paddlewheels. Later, propellers became a more common method of propulsion. A propeller works

◀ *Large, 19-foot (5.75-meter) diameter propellers have been fitted to the cruise ship* Millennium. *These propellers are attached to electric propulsor units, connected to an electrical generator inside the ship. A recent development, electrical propulsors are more mechanically efficient than other forms of propulsion and are cheaper to maintain.*

by "screwing" through the water and forcing it backward. The backward movement of the water pushes the propeller forward, and this movement is sent to the hull of the boat through a fitting called a thrust block. A more recent method of propulsion, developed by New Zealand engineer William Hamilton (1899–1978) in the 1950s, is the waterjet. This propulsion method uses a powerful jet of water to move the boat forward. A waterjet generates propulsive thrust because of the reaction created when the water is forced backward—the same as the thrust felt when holding a powerful hose. The discharge of the high-speed water jet generates a reaction force that is transferred through the body of the jet unit to the boat's hull, propelling it forward.

Outboard motors

An outboard motor is a small, self-contained engine used mostly to power smaller boats. It is usually fixed to the outside of the boat at the back

▲ *This photograph shows a rescue boat with an outboard motor. The propeller is housed inside a metal cage so that it cannot injure people in the water.*

and is therefore called an "outboard" motor. It can be unclamped from the boat and removed for repair or for storing during the winter.

The earliest outboard motor was the French Motogodille, which appeared in 1902. Other models soon followed, but Norwegian-born U.S. engineer Ole Evinrude (1877–1934) is generally credited with designing and building the world's first commercially successful outboard in 1909, after about three years of trials and testing.

The powerhead

Outboard motors consist of two parts—the engine and the propeller—which are joined by a shaft. The part with the engine is called the powerhead and is at the top of the motor, out of the water. The powerhead also houses the electrical system, and on smaller motors, the fuel tank.

Nearly all outboard motors have a two-stroke gasoline engine, though a few have four-stroke or even electric motors. Rotary engines have also been developed for use in outboard motors.

The engine itself is like a small automobile engine mounted on its end. The cylinders are horizontal and drive a vertical crankshaft and driveshaft. Most outboards are water-cooled. The water is sucked in near the propeller and pumped up beside the driveshaft to the engine. There, it is circulated around the hot engine parts. The water is then taken back down the shaft and is discharged just above the propeller. The engine exhaust is also taken down the shaft from the exhaust ports and is discharged near the propeller.

The lower unit

The lower unit holds the horizontal propeller shaft, the pump for pushing the cooling water up to the engine, and (except on the smallest motors) a gearbox for forward, reverse, and neutral. The spinning propeller tends to push the stern to one side, so the lower unit has a small fixed rudder, called the trim rudder, to keep the boat moving in a straight line.

Control

Outboards are steered by turning the whole motor, including the propeller. Small engines have a tiller that often has another lever mounted on it for controlling the speed. The powerhead also has a gear lever on it if there is a gearbox.

Larger outboards would be too powerful to steer with a tiller, so they are steered by a wheel linked to the motor by cables. More than one motor is often installed for extra power and speed. In this case, the motors are always linked together by cables and are steered using a wheel.

Water jets

The most modern method of marine propulsion is the waterjet drive. This consists of a pump that draws water in from beneath the boat's hull and forces it out through a nozzle at the stern. This form of propulsion is used by some hydrofoils and

▲ *Waterjet propulsion devices can have reversing and steering systems or provide forward thrust only. They are a very powerful and effective method of propulsion and are used on many fast craft.*

many fast craft. Since no part of a waterjet unit sits below the hull of the boat, these propulsion devices are particularly useful in shallow water.

The jet unit is mounted inboard in the aft (rear) section of the boat's hull. Water enters the jet unit through an intake on the bottom of the boat at whatever speed the boat is traveling. The water is pumped through the jet unit and accelerated, and then is discharged through a nozzle at the rear of the boat at a high speed.

The pumping unit is driven by an engine (usually diesel or gasoline), connected by a driveshaft. The pumping is done by an impeller. An impeller is similar to a propeller, but instead of forcing water backward, it draws in water. The pressure of the water flow is increased further as the water is forced through a tapered tube and is discharged through the nozzle at the rear at high speed.

Steering is achieved by changing the direction of the stream of water as it leaves the jet unit. Pointing the jet stream one way forces the stern of the boat in the opposite direction, and this steers the vessel into a turn.

See also: DIESEL ENGINE • GAS TURBINE • HYDROFOIL • INTERNAL COMBUSTION ENGINE • PROPELLER • SAILING • SHIP AND SHIPBUILDING • STEAM ENGINE • SUBMARINE

Mars

Mars is the fourth planet from the Sun and Earth's nearest neighbor after Venus. Out of all the planets in the solar system, Mars most resembles Earth. Like Earth, Mars is made of solid rock, and it is the only planet that has an atmosphere or daytime temperature anything like Earth's. Even Martian days last almost as long as Earth days. Long ago, space probes confirmed that the surface is a lifeless and dry desert. However, recent missions to Mars confirmed that the planet once had oceans and rivers. Scientists hope soon to find traces of ancient life in Martian rocks. Within the next 20 years or so, Mars may become the first planet beyond Earth to be visited by people.

Mars is the seventh largest planet, 4,216 miles (6,786 kilometers) in diameter—a little more than half the size of Earth. It revolves around the Sun at an average distance of some 142 million miles (228 million kilometers) and takes 687 Earth days to make one revolution (the Martian "year").

To the naked eye, Mars is just a point of light in the night sky. Even through powerful binoculars, it is nothing more than just a pink blob. Mars is only clearly visible for a few months or so every two years when it is directly opposite us, away from the Sun. It is then said to be "in opposition." At this point, it is just 35 million miles (56 million kilometers) away and appears almost as bright as Jupiter in the night sky.

▶ *Mars's red-ocher surface is clearly visible in this photograph. So, too, is the white polar ice cap, white clouds, and the dark patches once thought to be seas but now known to be patches of shifting sand.*

Astronomers can see Mars best when the planet is in opposition to Earth. Through a telescope, Mars appears red, which earned it the nickname the "Red Planet"—and also its real name Mars, the bloody ancient Roman god of war. About one-third of the planet is covered with faint markings called maria (*singular,* mare). *Maria* means "sea," but astronomers now think that the maria were never covered by water. Mars's two tiny moons, Phobos and Deimos, can also be seen using a powerful telescope. They are just small, cratered lumps of rock. Even the larger of the two, Phobos, is no more than 15 miles (25 kilometers) across.

Martian weather

The Martian atmosphere is very thin—about one-hundredth as dense as the atmosphere of Earth. There are signs that it was denser in the past and was once filled with clouds of water vapor. But it has not rained on Mars for millions of years, clouds are sparse, and the atmosphere now consists mostly of carbon dioxide gas (CO_2), with minute traces of other gases, such as nitrogen (N_2), oxygen (O_2), and carbon monoxide (CO).

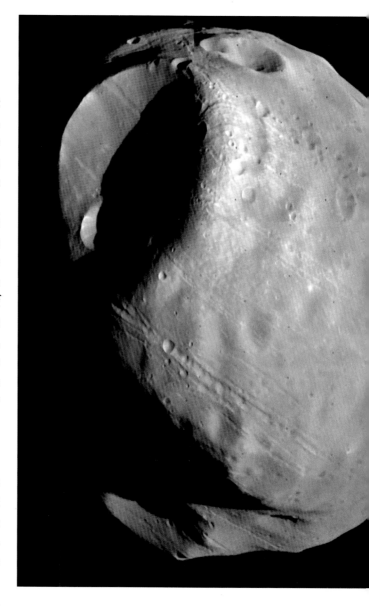

▶ *This composite view of Phobos was created from photographs sent back by the Viking I space probe. The giant crater on the left is Stickney crater, named for the wife of the moon's discoverer.*

Over much of the planet, temperatures may rise at noon in midsummer to as high as 50°F (10°C). At night, however, the temperature falls rapidly to –110°F (–80°C) or below, because the Martian atmosphere is so thin that it cannot insulate the planet against heat loss.

At the poles in midwinter, the temperature can plummet to –220°F (–140°C). The carbon dioxide freezes out of the atmosphere as dry ice, snowing to give Mars polar ice caps that resemble Earth's ice caps. These ice caps also contain large amounts of water in the form of ice. The ice caps vary in size from season to season. In the summer they shrink, and in the winter they grow. The ice cap at the southern hemisphere nearly disappears during the southern summer on Mars, which is slightly hotter than the summer in the northern hemisphere.

The Martian surface

The photographs sent back by various space probes and landing craft have shown a variety of different features on the surface of Mars. Indeed, space probes orbiting the planet have now scanned Mars's surface so precisely that it is possible to draw maps of Mars every bit as detailed as maps of Earth.

Much of the landscape consists of rust-colored sandy desert, strewn with rocks. The southern hemisphere of Mars is heavily cratered and has two

DID YOU KNOW?

Mars's two moons were discovered on August 17, 1877, by U.S. naval observer and astronomer Asaph Hall (1829–1907)—but only because his wife insisted that he continue looking through his telescope one night when he wanted to pack up and go to bed. In honor of his wife, Hall named a giant crater on Phobos for his wife's maiden name—Stickney.

massive basins, probably caused by the impact of giant meteorites. The basins are called Argyre, about 500 miles (800 kilometers) across, and Hellas, which is twice as big.

In the northern hemisphere of Mars, there is a vast volcanic plateau dominated by four enormous volcanoes. The largest, Olympus Mons, is about 400 miles (600 kilometers) across, making it by far the largest volcano in the solar system and ten times as big as Hawaii's Mauna Kea, which is Earth's largest volcano. Olympus Mons soars more than 15 miles (25 kilometers) into the Martian sky, making it three times as tall as Mount Everest. At places on its slopes are cliffs more than 4 miles (7 kilometers) high.

To the east of Olympus Mons is a Martian "Grand Canyon" called Valles Marineris. It is a great gash stretching around the planet for 3,000 miles (5,000 kilometers) and, at places, it is 250 miles (400 kilometers) wide and 6 miles (10 kilometers) deep. Recent space probes have revealed island-shaped mesas, mountain ridges, and valleys along the floor of this valley. Scientists believe these may have been carved by glaciers millions of years ago.

Changing ideas about Mars

Back in the 1880s, many scientists thought Mars may actually be home to intelligent life after U.S. astronomer Percival Lowell (1855–1916) spotted dark lines on the planet's surface through his telescope. Lowell believed they were canals built by

▼ *These false-color views of the surface of Mars were compiled from data sent back by the Mars Orbiter Laser Altimeter (MOLA), an instrument aboard NASA's* **Mars Global Surveyor.**

Martians. In the 1950s, powerful telescopes showed that these "canals" were simply optical illusions. For a century, however, some astronomers had hoped they might find at least primitive life on Mars.

When the Viking space probes landed on Mars in the 1970s, it showed there were no signs of microscopic life. The Viking experiments seemed to show that Mars is lifeless and always has been.

In 1996, the hunt for life on Mars began again. Every now and then, meteorites crash into Mars so hard that chunks of the planet fly off into space and end up on Earth. Scientists from the National Aeronautics and Space Administration (NASA) analyzed one of these rocks and found microscopic rods they said could only be made by living organisms. Many scientists disagreed, but then bacteria were discovered living underground on Earth in extreme conditions. So there may be signs of life on Mars after all, buried in rocks where the Viking probes could not find it.

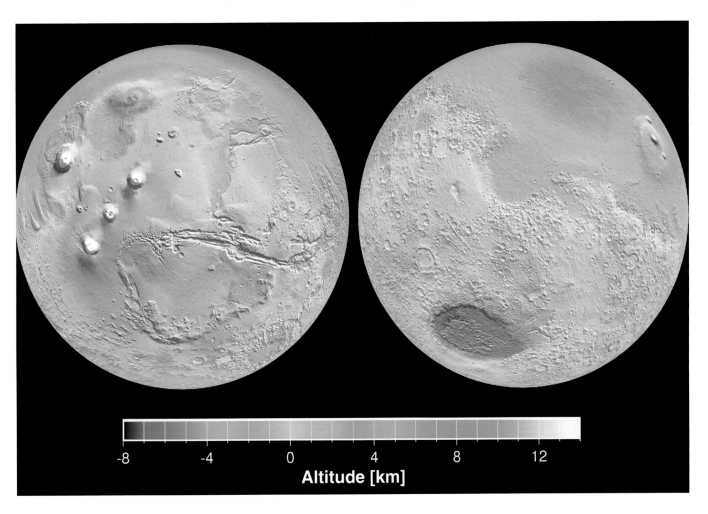

◄ *This is a picture of the* **Opportunity** *robot vehicle, which roamed across the Martian surface early in 2004 and proved conclusively that Mars was once a wet planet. The large mainmast extending upward carries a panoramic camera and a temperature probe. The hinged arm at the front of the vehicle carries a rock scraper that can be extended to study Martian rocks.*

Water on Mars

The early twenty-first century has seen a string of space missions to Mars. The earliest missions have all been unpiloted explorations of the planet. However, both NASA and the European Space Agency (ESA) are now planning crewed missions within 30 years. ESA's Aurora program plans to bring back rock samples from Mars in 2014 and put astronauts on the planet by 2033. NASA hopes to bring back rock samples from Mars as early as 2005.

Early in 2004, three missions to Mars began to send back data that is transforming astronomers' views of the planet—ESA's *Mars Express* and NASA's *Spirit* and *Opportunity*. It was hoped that the *Mars Express* would land a probe called *Beagle* on the planet, which would detect microscopic life below the surface. *Beagle* was lost, but *Mars Express* beamed back data that showed that Mars's atmosphere contains methane (CH_4), often taken as a sign of life. *Mars Express* also confirmed that there was water ice in Mars's ice caps.

NASA's twin missions, *Spirit* and *Opportunity*, both touched down successfully and sent out robot explorer vehicles. Astonishingly, both found abundant evidence that Mars was clearly a watery planet in the past. *Spirit* landed in the Gusev crater, which seemed to be the bed of an ancient lake. Meanwhile, *Opportunity* landed in Meridiani Planum, and there were signs that this was once the shore of a shallow, salty sea. Unmistakable ripple marks were clearly visible in the images sent back to NASA scientists on Earth. So were grains of sand shaped by running water, not wind. There was even evidence of sedimentary rocks—rocks created from sediments settling in water.

Scientists now have no doubt that Mars was once a watery planet, washed with rivers, lakes, and oceans, before it was dried out by a runaway greenhouse effect. If it was once so wet, there seems every chance that it developed life in the same way as life developed on Earth. Scientists are already talking about the possibility of finding Martian fossils. And if there was life there once, there may be life there now—not just below ground but actually on the surface.

See also: SOLAR SYSTEM • SPACE PROBE • SUN

Mass and weight

Mass and weight are different ways of describing how much matter an object contains. Weight is a measure of the gravitational force on an object and is proportional to the object's mass and the masses of objects exerting gravitational forces on it.

People usually use the word *weight* when talking about how heavy something is. They measure weight in units such as pounds and kilograms. But astronomers and physicists prefer to talk about mass. The distinction is important. When scientists talk about mass, they are talking about how much matter it contains. Something that is heavy contains a lot of matter and is said to be high in mass. Something that is light contains only a little matter and is said to be low in mass.

Mass versus weight

Mass is related to a property called inertia. If a person throws a pebble from one hand to the other, he or she will notice that it takes a push to get it going and another to stop it. If the pebble is at rest, it will not move unless it is pushed. If the pebble is moving, it continues moving until it is stopped. Inertia is this reluctance to change, and it depends entirely on mass. The heavier an object is, the more force a person will need to get it moving or to stop it from moving. So mass is a measure of how much inertia an object has, or a measure of the force needed to make it move or stop.

For scientists, weight is something rather different. Every object in the universe with mass—that is, every bit of matter—is drawn toward every other bit of matter by the force of gravity. The power of the attraction depends on how massive the objects are and how far apart they are. The pull between relatively small objects, such as oranges, is so small it is barely noticeable. However, the pull between an orange and a giant object such as Earth—the pull that makes the orange fall to the ground when dropped from a height—is so large it can be measured very simply with weighing scales. For scientists, weight is therefore a measure of the force of gravity.

Gravity and acceleration

Gravity works by making things move faster together. This is called acceleration. If something falls, it gains speed until it hits the ground. Gravity makes everything accelerate at the same rate, no matter how heavy it is. Italian scientist Galileo Galilei (1564–1642) proved this with an experiment in which he dropped lead shot balls of different sizes and masses from the Leaning Tower of Pisa in Italy. All the balls reached the ground at the same time.

▶ *Pan balances measure what most people call weight, which scientists call mass, by counterbalancing a mass identical to the object being measured. Spring balances measure what scientists define as weight.*

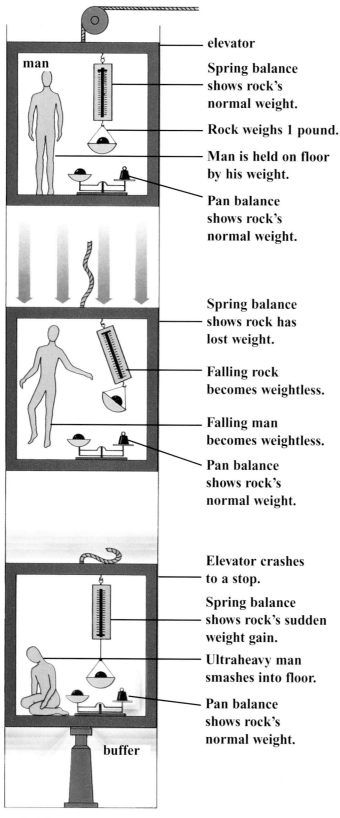

elevator

man

Spring balance
shows rock's
normal weight.

Rock weighs 1 pound.

Man is held on floor
by his weight.

Pan balance
shows rock's
normal weight.

Spring balance
shows rock has
lost weight.

Falling rock
becomes weightless.

Falling man
becomes weightless.

Pan balance
shows rock's
normal weight.

Elevator crashes
to a stop.

Spring balance
shows rock's sudden
weight gain.

Ultraheavy man
smashes into floor.

Pan balance
shows rock's
normal weight.

buffer

▲ *A falling elevator shows how, unlike mass, weight
is a force that can be gained or lost. As the elevator
plunges toward the buffer, the man and rock become
weightless. As the elevator hits the ground, they both
temporarily gain an enormous amount of weight.*

DID YOU KNOW?

Objects weigh less on the Moon than they do on Earth simply because the gravity of the Moon is six times weaker than that of Earth. But their mass remains the same.

Fixed mass, varied weight

An object always has the same mass, no matter where it is, because it always contains the same amount of matter and so always has the same inertia. However, weight varies from place to place. A large, heavy planet, for example, is more massive than a smaller, lighter planet. So the gravity of the large, heavy planet is greater, and everything weighs more there. Someone on the giant planet Jupiter would find it hard to even lift a finger. Someone on the small planet Pluto, however, would easily be able to throw a table into the air.

Weightlessness

There are places where weight effectively disappears altogether. Astronauts orbiting Earth in spacecraft experience weightlessness, when everything floats around the cabin as if it weighed nothing at all. They have lost none of their mass, however. If an astronaut is hit by a flying chair, it will still hurt just as much. Nor have they escaped from Earth's gravity, although this might seem to be so. Weight is the acceleration due to gravity, and what is actually happening is that the spacecraft is moving around the Earth so fast that the acceleration (and thus weight) is effectively canceled out.

Falling elevator

To see how this works, imagine an elevator plunging down the elevator shaft when the cable breaks. As the elevator falls, a man standing inside will start to float off the floor. The man and the elevator are both falling at much the same rate, since gravity accelerates everything equally. Although their mass is unchanged, they both become effectively weightless as they plunge down the shaft, accelerated equally by gravity.

This weightlessness can be measured. Imagine a spring balance hanging from the elevator's ceiling. The balance is weighing a rock. Before the elevator starts to fall, the spring balance shows that the rock weighs 1 pound (0.45 kilograms). As the elevator falls, the stone floats upward, and stops pulling on the spring, effectively becoming weightless. A pan balance will not show this, because the weights in the pan lose their weight in the same way as the rock, so there is no change.

As soon as the elevator hits the bottom of the shaft and comes to a stop, everything instantly regains its weight. In fact, the acceleration of the fall has added tremendously to the weight of everything in the elevator. So when the elevator hits the ground, the man smashes into the floor as if he weighed a huge amount. The effect is very brief, because the acceleration is quickly brought to a halt by the impact, but again it is measurable. Indeed, the weight gain is clear from the spring balance, which is tugged so hard by the still falling stone that its weight briefly registers as many times more pounds than the original measurement.

Einstein's mass

Scientists once thought that mass is totally unchangeable and fixed throughout the universe. This is still true in most cases. However, German-born U.S. physicist Albert Einstein (1879–1955) changed that simple notion of mass in 1905 when he published his special theory of relativity. In fact, it now seems that mass is actually interchangeable with energy, so the mass of an object can actually change as the energy changes.

This all means very little in normal situations. As an object begins to travel very fast, however, at speeds approaching the speed of light, its mass increases dramatically. Scientists have to think in terms of a rest mass, which an object has when it is theoretically still, and the mass gain when it begins to move. This extra mass becomes crucial when scientists start to release the energy of atoms during nuclear reactions.

See also: CENTRIFUGE • EINSTEIN, ALBERT • GRAVITY • PHYSICS • RELATIVITY

Mathematics

Mathematics is a vital part of human knowledge. By making calculations and logical progressions with numbers and symbols, mathematics enables people to do everything from figuring out shopping bills or doing their monthly budgets to more complex tasks, such as predicting the flight of a spacecraft or even demonstrating the entire history of the universe.

People were using numbers to count thousands of years ago. In fact, even small animals have a basic number sense. However, it was probably only when people settled down to farm, around 10,000 years ago, that they began to think of large numbers. Early farmers needed numbers to count cattle or figure out how many bags of wheat to take to market. Written numbers first appeared in the Middle East, along with the first farms and towns.

Early people may have counted on their fingers. This works well but leaves no record of the count. So people began to make number records in ways such as dropping stones, shells, or clay disks into a bag. Then in Sumeria, about 6,000 years ago, they started to record numbers by scratching on clay tablets—the first written numbers. The ancient Babylonians used different marks to indicate different large numbers, which is much the same method people use today.

These ancient civilizations learned not just to count and write numbers down, but they developed surprisingly sophisticated mathematical skills, too. More than 5,000 years ago, the Babylonians were adept in arithmetic—the art of figuring things out using numbers. Babylonian schoolchildren learned how to multiply and divide, using arithmetical tables to help with harder calculations. The great Babylonian mathematicians could solve complex equations.

▲ Greek mathematician Euclid laid down the basic principles of geometry in his book **The Elements**, which he wrote about 2,300 years ago in Alexandria in present-day Egypt.

Geometry and the Greeks

Geometry was also well developed in ancient times. Geometry is the mathematics of shapes and was probably first used to help people figure out the area of the land they owned. It was developed to a sophisticated level by the ancient Egyptians, who used geometry to help them build perfect pyramids. In 1858, Scottish Egyptologist Alexander Henry Rhind (1833–1863) found a papyrus (a document written on dried reeds) by the Egyptian scribe Ahmes around 1650 BCE. The Rhind papyrus showed that the Egyptians knew much about the geometry of triangles. For example, they knew how to figure out the height of a pyramid from the length of the shadow it cast.

However, the first great masters of geometry were the ancient Greeks. *Geometry* is a Greek word meaning "Earth measurement." Greek scholars such as Pythagoras (c. 580–c. 500 BCE), Eudoxus of Cnidus (c. 400–c. 350 BCE), Appolonius of Perga (c. 262–c. 190 BCE), and many others made major contributions. But the greatest contributor of them all was Euclid (lived c. 300 BCE). Euclid's book, entitled *The Elements,* was such a brilliantly thorough study of geometry that it became the framework for geometry for thousands of years. Even today, mathematicians still refer to the geometry of flat surfaces—lines, points, shapes, and solids—as Euclidean geometry.

Merchants and others in the ancient world used math for its practical benefit (now called applied math), but the Greeks developed purely theoretical math (now called pure math). Starting with Thales of Miletus (c. 625–c. 547 BCE), they made mathematics a logical system. The Greeks introduced the idea of proofs and the idea that rules could be figured out logically from certain assumptions, or postulates, such as "A straight line is the shortest distance between two points." Assumptions are combined to make a basic idea for a rule, called a theorem, which then must be proved or disproved.

Mathematics from the East

After the collapse of ancient Greek and Roman civilizations, the next major advances in math came from the East—from the great Arabic and Indian mathematicians, such as Nasir ad-Din at-Tusi (1201–1274 CE). A key contribution was a book written about 825 CE by Arab mathematician Muhammad ibn Musa al-Khwarizmi (c. 780–c. 850 CE). In his book, he described the decimal number system developed in Hindu India. This number system was much more versatile than earlier systems and remains the basis of the number system in use today. Roman numbers could only make larger numbers by adding more digits or

▶ *Algebraic calculations carried out by these schoolchildren were developed by Arab scholars more than one thousand years ago. These calculations are now central to most mathematical operations.*

different symbols. In the Arabic-Hindu system, the position of a digit alters value tremendously. A 2 by itself is small, but a 2 before 5 is much bigger. In this way, the Arabic-Hindu system could repeat the same basic symbols again and show huge numbers.

The same Arab-Hindu system also gave us the idea of 0, or zero, which has become fundamental to modern mathematics. Al-Khwarizmi also wrote a book about algebra—indeed, the word *algebra* comes from the title of his book. In the mid-1100s, the Arabic-Hindu number system was introduced to Europe and replaced the Roman numerals that had been in use for more than one thousand years. In 1202, Italian mathematician Leonardo Fibonacci (c. 1170–c. 1240 CE) introduced algebra to Europe, and modern mathematics began to develop.

Mathematics and science

Another crucial development was the invention of logarithms by Scottish mathematician John Napier (1550–1617) in 1614. Logarithms reduce even the most complex calculations to simple additions of decimal fractions. They made practical the complex calculations involved in astronomy. With tools like these, French mathematician and philosopher René Descartes (1596–1650) created analytical geometry in 1637, which added the power of algebra to geometry and became crucial for calculations involving variables such as forces.

◄ *René Descartes developed analytic, or Cartesian, geometry to translate geometric problems into algebraic form so they can be solved using equations.*

non-Euclidean geometry. German mathematician August Möbius (1790–1868) developed topology, a mathematical method of studying the way in which a surface bends or stretches, and French mathematician Jules Poincaré (1854–1912) pioneered probability theory.

BRANCHES OF MATHEMATICS

Mathematics has many branches. Each one uses different methods and works on different kinds of problems. Arithmetic, geometry, algebra, and calculus are the main branches. Other branches include trigonometry and analytic geometry, probability and statistics, and set theory and logic.

Arithmetic

Arithmetic is at the heart of all mathematics. It is the art of figuring things out using numbers and is the oldest of all mathematical skills. It is founded on four basic operations: addition, subtraction, multiplication, and division.

Each of these operations is related and is just a way of counting. Addition is repeated counting forward, adding numbers together to make bigger numbers. Subtraction is repeated counting backward. Multiplication is just a quick way of doing repeated addition, and division is a quick way of doing repeated subtraction. Together, these four operations form the basis of all mathematics.

Addition means putting two numbers together— known by mathematicians as addends—to get a third number called the sum. Put another way, addition means increasing one number, or addend, by another, the augend. It is basically a process of piling up or counting. For example, imagine a pile of 12 socks (the addend), add six more (the augend), and count how many socks there are altogether (the sum; in this case, 18). Subtraction is the opposite of addition and involves taking one number (the subtrahend) away from another (the minuend) to get the answer.

A few years later, English physicist and mathematician Issac Newton (1642–1727) and German mathematician and philosopher Gottfried Leibniz (1646–1716) developed calculus. Although calculus is little used in everyday calculations, it underpins nearly all scientific calculations. Over the next century, it was developed into an immensely powerful tool by innovators such as Swiss mathematicians and brothers Jakob Bernoulli (1654–1705) and Johann Bernoulli (1667–1748), Swiss mathematician Leonhard Euler (1707–1783), and Italian-born French mathematician Joseph Lagrange (1736–1813), who showed that, like analytic geometry, it could work equally well with both algebra and geometry.

By the 1800s, mathematicians were exploring entirely abstract, complex ideas that at the time seemed to have little to do with the real world—but have now proved much more relevant than perhaps they ever imagined. German scientist Carl Gauss (1777–1855), Russian mathematician Nikolay Lobachevksy (1792–1856), and Hungarian mathematician János Bolyai (1802–1860) created a

Mental arithmetic

When mathematicians first learned to make quick mathematical calculations, it seemed like magic to many people. In fact, ancient Chinese arithmetic processes seemed so tricky and clever that they were still being used by Chinese "mind-readers" in the music halls of the nineteenth century.

Most people use electronic calculators to find sums, but mental arithmetic—calculating the answer in one's head—is still useful. The key to effective mental arithmetic is to simplify the sum to create an easy calculation. Multiplying by ten is easy enough—simply add 0 to the end of the other number. Other numbers can be simplified, too. One useful trick is to round numbers up to ten and then multiply by ten. To multiply 38 by 17, for example, round the 38 up to 40 (subtract the extra two 17s at the end of the calculation). Take the 0 off the 40 and multiply 17 by 4 to give 68. Now return the 0 to give 680. Finally, take off the extra two 17s, which makes 34, to give the answer 646.

Geometry

Geometry is the mathematics of regular shapes. It is about points, lines, angles, triangles, circles, squares, solids—every shape one can imagine and many more besides. Many regular shapes occur naturally, from the symmetry of crystals to the perfect hexagons (six-sided shapes) of some honeycombs. Many shapes are used by people to build bridges and houses and appear in everything from analyzing the structure of molecules to programming the flight path of a satellite.

Basic geometry is essentially about lines and the angles between them and how they make up two basic kinds of shapes—the circle and the polygon. Polygons are flat shapes with straight sides. They can have any number of sides, from three to infinity, and the sides can be any length. Triangles, squares, rectangles, and hexagons are all polygons.

Mathematicians analyze shapes or "figures" in a particular way. They might try to calculate the area of a triangle, for example, or figure out the relationship between particular angles. Or they might try to discover what special properties a shape has. It is known, for example, that a square has four sides of equal length at right angles to each other, and that the diagonals of a square—the lines that cut across from one corner to the other—always cross exactly in the middle. How many other properties might a square have?

In recent years, an entirely new kind of geometry has helped push astronomers' understanding of the universe to new limits. Geometry is the mathematical tool that allows scientists to create theoretical models of multidimensional space. For more than 2,000 years, the basic geometry of flat surfaces developed by Euclid seemed quite adequate for most purposes. But the exploration of multidimensional and curved space needs an entirely new kind of geometry, such as that pioneered by Gauss and Lobachevsky and developed by German mathematician Georg Bernhard Riemann (1826–1866) in the 1860s. Riemannian geometry is the geometry of a sphere, in which all straight lines are complete circumferences, that is, the edges of a circle.

◄ *Sophisticated calculators such as this model from Texas Instruments, have taken over much of the work of mathematical calculations, but the basic operations are still the same.*

Algebra

Algebra is a valuable form of mathematics that uses letters or symbols as well as numbers. The idea is that the letters or symbols represent unknown numbers. The unknown letters, typically a and b, or x, y, and z, are called variables, because they can represent any number. When mathematicians need to figure out what the unknown number is, they substitute it with a letter in an equation.

An equation is a way of saying that two things are equal to each other. So an equation always has two halves, with an equals sign (=) between them. A simple equation in arithmetic might be:

$$2 + 3 = 5$$

In algebra, 3 might be the unknown number, x, so the equation would be written:

$$2 + x = 5$$

Once an equation is written algebraically, the mathematician can treat the variable just like any other number to "solve" the equation and find out what the mystery number is.

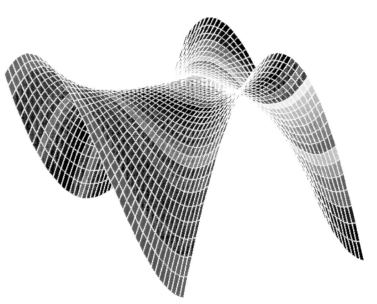

▲ *This a computer representation of a mathematical system for generating random numbers. It was created by Japanese mathematician Makoto Matsumoto (1965–). Called a "Mersenne Twister," it uses complex geometry based on 623 dimensions.*

Algebra has proved its worth for everyone from scientists to stockbrokers. It can be used to show the path a football will take, what electrical current is needed to make a lightbulb glow, or to discover just when the universe began.

Wherever there is an unknown quantity, algebra can help find out what it is or show how different quantities relate to each other. The area of a rectangle, for example, is its length multiplied by its width. If the area is A, the length l, and the width w, this can be expressed algebraically:

$$l \times w = A$$

Perhaps the best known algebraic expression of all time is the equation of mass-energy equivalence derived by German-born U.S. physicist Albert Einstein (1879–1955):

$$E = mc^2$$

In this, E stands for energy and m for mass. The c is the speed of light (186,000 miles, or 300,000 kilometers, per second). It stands for *celer*, a Latin word meaning "fast." As Einstein showed, the speed of light is the same everywhere in the universe. This equation helped scientists unlock the energy of the atomic nucleus and create the atomic bomb.

Calculus

Calculus is one of the most important of all mathematical discoveries. Although its origins date back to the ancient Greeks, it was officially developed independently by Newton and Leibniz.

Calculus is the branch of mathematics that deals with rates of change; that is, how fast things are changing. A rate of change might be anything from how fast a car is accelerating from the traffic lights to how quickly a cyclist can change direction as she rounds a bend on her bicycle. What makes calculus so astonishing is that it means all processes can be treated mathematically. Newton used it to show why the planets follow elliptical orbits around the Sun as they are held in balance by their own momentum and the pull of the Sun's gravity.

▲ *The ASCI White computer is 16,000 times more powerful than a desktop PC. It can perform as many calculations in one second as it would take a person with a calculator ten million years to do.*

In fact, calculus can be used to make a mathematical model of anything that moves or changes, which means everything that ever happens. It can be used to calculate anything from the flight of a baseball to the way a magnetic field changes. No major advance in most branches of science would have been possible without calculus. Nor could scientists have ever figured out the trajectories needed to launch vehicles into space.

How calculus works

There are two kinds of calculus. Differential calculus helps mathematicians figure out how fast something is changing. Integral calculus helps mathematicians figure out how it is changing (for example, where or how big the change is) any time or place if the mathematician already knows how fast it is changing. Integral calculus is especially useful for calculating areas and volumes.

Two things are central to calculus—functions and variables. A function is something that is the way it is because of something else. The area of a square, for example, depends on how long the sides are, so the area is a function of the side length. How far a cyclist travels when riding his bike at a certain speed depends on how long he has been riding, so the distance is a function of the time.

A variable is a quantity that changes. If speed is constant, it does not change, so it is not a variable. Time and distance do change, so they are variables. One variable is usually dependent on the other. The distance the cyclist travels depends on the time he has been riding. Distance is a dependent variable. Time is an independent variable—it changes regardless of the distance the cyclist rides.

A change can be a change in direction, such as a bend, a curve, or a circle. Calculus is concerned with just how fast direction is changing. The direction at any point is the tangent—a straight line drawn across the edge of the curve. So calculus can be the math of circles, curves, and tangents.

Calculus is linked to the math of graphs, also called analytic geometry, because a rate of change can be plotted as a curve (line) on a graph. Typically, the independent variable, such as time, is plotted on the *x*-axis horizontally across the graph. The dependent variable, such as distance, is plotted on the *y*-axis vertically up the graph. Here calculus overlaps with trigonometry, which is the branch of math that deals with circles, triangles, oscillations, and waves. Using right-angled triangles, such as a straight line on a graph, and its coordinates, trigonometry enables calculations that describe angles, turning, and swinging.

See also: ALGEBRA • CALCULUS • MAPMAKING • NUMBER SYSTEM • PHYSICS

Maxwell, James Clerk

Scottish physicist James Clerk Maxwell is best known for his studies of electromagnetism. He was the first to prove that light travels in the form of waves and that it is part of the electromagnetic spectrum. Maxwell also contributed to the study of gases and color vision.

James Clerk Maxwell was born in Edinburgh, Scotland, on November 13, 1831, and was educated at Edinburgh Academy between 1841 and 1847. Maxwell's career started from an early age. When he was 14, he read a paper on geometry at the Royal Society of Edinburgh. A year later, he demonstrated his analytical talent by devising an original method for drawing a perfect oval.

In 1847, Maxwell attended lectures at Edinburgh University, where he researched the properties of polarized light and the engineering problems of bending beams and twisting cylinders. In 1850, Maxwell went to Cambridge University to continue his research. There he made his first important contribution to physics with his work on color vision, extending the work of English physicist and physician Thomas Young (1773–1829) and German physicist and physiologist Hermann von Helmholtz (1821–1894). Maxwell figured out that all colors could be made using combinations of the three primary colors of light—red, green, and blue—and proved his theory by spinning disks on which he had printed colored sections of various sizes.

In 1854, Maxwell graduated from Cambridge University. A year later, he published the first of a series of papers on the electromagnetic theory of light. In the paper, entitled "On Faraday's Lines of Force," Maxwell built on the work of the great

English experimental physicist Michael Faraday (1791–1867). By comparing the behavior of the fields of lines of force that surround conductors and magnets to the flow of a fluid in a tube, Maxwell derived equations that represented known electric and magnetic effects.

In 1856, Maxwell was appointed professor of natural philosophy at Aberdeen University. Four years later, at the young age of 29, he became professor of natural philosophy and astronomy at Kings College, London. Maxwell was extremely productive at Kings. There, he did most of his work on the kinetic theory of gases, building on the achievements of German physicist Rudolf Clausius (1822–1888). Clausius had shown that gases consist

▶ *James Clerk Maxwell is depicted here in a photogravure from the 1870s, when he was a professor of experimental physics at Cambridge University.*

▲ *Between 1855 and 1859, Maxwell set about solving the problem of Saturn's rings. No one could figure out an explanation for the rings that would yield a stable structure. Maxwell calculated that the rings consisted of many small bodies in orbit around the planet.*

of molecules in constant motion, constantly colliding with each other and with the walls of their container. Maxwell developed what is now known as statistic mechanics to arrive at a formula to explain the distribution of velocity in gas molecules, relating it to temperature. In this way, Maxwell finally proved that heat resides in the motion of molecules.

The wave nature of light

Maxwell's greatest contribution to physics also came during his time at Kings College. In 1864, he published a paper in which he developed the fundamental equations that describe electro-magnetism. Maxwell showed that light is a form of energy that travels in two waves—one magnetic and one electric—which vibrate at right angles to each other and to the direction of motion. Maxwell summed up his work on the subject in *Treatise on Electricity and Magnetism,* published in 1873.

Later years

Following the death of his father in 1865, Maxwell went into semiretirement, returning to his family estate in Scotland. There, he continued his research on the kinetic theory of gases. In 1868, Austrian theoretical physicist Ludwig Boltzmann (1844–1906) modified Maxwell's earlier work on the kinetic theory of gases to explain heat conduction, resulting in the Maxwell-Boltzmann distribution law. Thereafter, Maxwell and Boltzmann both refined the kinetic theory of gases, making it applicable to all properties of gases. This work also prompted Maxwell to make an accurate estimate of the size of molecules and to devise a method of separating gases in a centrifuge.

In 1871, Maxwell was persuaded to become the first professor of experimental physics at Cambridge University. One of his main duties involved setting up the Cavendish Laboratory, which was opened in 1874. Maxwell continued in this position until 1879, when he fell seriously ill with cancer. He died in Cambridge the same year.

See also: ELECTROMAGNETISM • GAS • PHYSICS

Measurement

Measurements are essential both in many aspects of everyday life and in science. Sometimes only rough measurements are needed, for example, the time it takes a person to walk to work. Sometimes, though, scientists need to measure things with a far greater degree of accuracy.

There is often more than one way of measuring something. For example, a person could measure the length of his or her backyard by stretching a tape measure across it (probably having to do so several times to cover the whole distance). Alternatively, the person could first measure the length of his or her stride using the tape measure, then count how many strides it takes to walk the length of the yard and multiply the results together. It is often necessary to measure one thing as part of measuring something else. The same person could also use a surveyor's device, which is an instrument that emits pulses of ultrasound (sound with a pitch too high for human beings to hear) to measure distances. The time taken for an echo to return from a fence at the end of the backyard is measured by the instrument, which calculates the distance of the object and displays it to the user on a screen.

When surveyors want to measure longer distances, for example, several miles to a tower, they may use a method called triangulation. They aim a telescope at the tower and measure exactly the direction it lies in. Then they move the telescope to another position and point the telescope at the tower again. The tower lies in a slightly different direction because the point of observation has been moved. Knowing the distance between the two positions of observation, and the change in the apparent direction of the tower, they can figure out the distance to the tower.

◄ *Lasers project a beam so straight that they can be used to measure very slight movements over huge distances. In this picture, laser beams are being used to measure changes in the ground level, amounting to just a fraction of an inch, in the Long Valley Caldera, California. The small change in ground level is due to volcanic activity.*

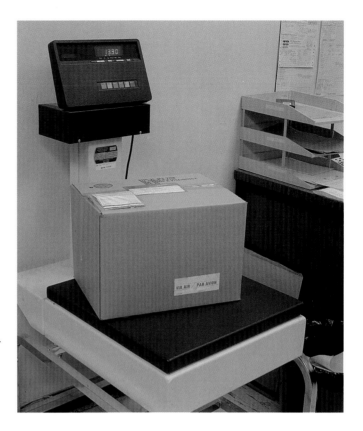

▶ *There is no need for precision in measurements taken by these digital mail room scales. The scales show the weight of the package to within 10 grams (about one-third of an ounce). This package weighs 13.90 kilograms (30 pounds and 11 ounces).*

Triangulation can be applied over the enormous distances of space. As Earth circles the Sun, nearby stars seem to move slightly compared to stars farther away. The nearer the star, the more it seems to move. By measuring how much it moves against the background stars, astronomers can figure out how far away it is. This method of measuring distance is called parallax. But it only works for nearby stars. For more distant stars, astronomers compare the color and brightness of a star. They expect stars of a certain color to be a certain brightness. For other distant galaxies, an array of other techniques can be used.

Precision and accuracy

Measurements can be more or less precise. No measurement is exact—it is simply more or less precise. The more precise a measurement is, the narrower are the limits it can vary on either side. Suppose a digital balance shows a cup of flour weighing 82 grams. For a balance that weighs to the nearest gram, the next measurement above would be 83 grams; the one below would be 81 grams. A more precise digital balance—one that weighs to the nearest one-twentieth of a gram—shows the cup weighs 82.4 grams. The next measurement above would be 83.45 grams; the one below would be 82.35 grams. The first measurement of 82 grams

is not wrong—it is simply less precise than the second. Another digital balance might be even more precise, showing the weight to the nearest thousandth of a gram.

Accuracy is the correctness of a measurement. A measurement might be very precise but simply wrong. For example, a kitchen spring balance might show the weight of some sugar as 103 grams. If the scale was not reading zero before the sugar was added, however, the weight would be wrong, even if the scale was extremely precise.

Errors of measurement

One of the most important advances in the progress of science has been the care scientists take not only to measure accurately, but to figure out clearly the degree of error they could be making. For example, the average distance of the Sun from Earth is known to within three miles. So an astronomer would state it as 92,955,832 ± (plus or minus) 3 miles (149,597,910 ± 5 kilometers).

When someone takes a measurement, there will almost always be some random and unpredictable error. Random errors are fairly easy to deal with.

> ## DID YOU KNOW?
>
> The first measurement of the size of Earth was made in the third century BCE. Greek scientist Eratosthenes (c. 276–c. 194 BCE) combined observations of the Sun made at two places in Egypt to find a size for the circumference of Earth. His early calculation was very close to the true value of 25,000 miles (40,000 kilometers).

The random errors tend to cancel out if the measurer takes many measurements of the same thing. Systematic errors are more difficult to deal with. These are errors that make the result always too big or always too small. For example, a ruler that is supposed to be 12 inches (30.48 centimeters) long may be ⅟₁₆ inch (0.159 centimeters) short.

Standards

To be certain exactly what each measurement is, international organizations use standard measures. When the meter was introduced in 1791, for example, the French government made a platinum bar exactly a meter long to be kept safe and act as the standard. Since then, various different standards have been adopted, each increasingly

▲ *The kilogram is the last base unit of the SI system that has a solid object as its standard. There are six copies of this standard kilogram, made of an alloy 90 percent platinum and 10 percent iridium. Each standard kilogram is kept in a vacuum.*

SI NUMBER PREFIXES

Example: The volt is an SI electrical unit with symbol V. Using the prefix *milli-* scientists make the millivolt, a thousandth of a volt, with the symbol mV. Using the prefix *kilo-*, they also make the kilovolt, a thousand volts, with the symbol kV.

yotta (Y)	=	1,000,000,000,000,000,000,000,000
zetta (Z)	=	1,000,000,000,000,000,000,000
exa (E)	=	1,000,000,000,000,000,000
peta (P)	=	1,000,000,000,000,000
tera (T)	=	1,000,000,000,000
giga (G)	=	1,000,000,000
mega (M)	=	1,000,000
kilo (k)	=	1,000
hecto (h)	=	100
deca (da)	=	10
deci (d)	=	0.1
centi (c)	=	0.01
milli (m)	=	0.001
micro (µ)	=	0.000001 (*µ* is the Greek letter *mu*)
nano (n)	=	0.000000001
pico (p)	=	0.000000000001
femto (f)	=	0.000000000000001
atto (a)	=	0.000000000000000001
zepto (z)	=	0.000000000000000000001
yocto (y)	=	0.000000000000000000000001

precise as science has demanded. Eventually, it was realized that no solid bar could be nearly precise enough. In 1983, the standard length of a meter was defined as $\frac{1}{299,792,458}$ the distance light travels in a vacuum in one second. This is measured using a helium-neon laser beam.

Measures of time, too, have been defined with increasing precision. The most accurate clocks now available are atomic clocks, which are accurate to within billionths of a second. Just as a drum vibrates at a certain frequency, so do atoms, but many billions of times in a second. An atomic clock works by responding to the electromagnetic radiation caused by atomic vibrations. Since 1967, one second has been defined as 9,192,631,700 vibrations of a cesium-133 atom. In the near future, the standard definition of a second will be given by light vibrating 100,000 times faster—at one quadrillion times a second.

Systems of units

Every measurement must be given in certain units. Many systems of units have been used in different countries. For example, people have measured distance in the past in such units as the rod, equal to 5.5 yards (5.029 meters).

In many cases, the same name has been given to units of different sizes.

For example, the troy ounce is slightly larger than the avoirdupois ounce. And the gallon used in Britain and other countries is 20 percent larger than the gallon used in the United States.

Scientists avoid such confusion by using their own special units, called SI units. "SI" stands for *Système International,* French words meaning "international system." There are seven basic SI units, including the meter for length, the second for time, the ampere for electrical current, and the kelvin for temperature. "Derived" units are combinations of the basic SI units and are usually named for famous scientists. For example, the pascal is the unit of pressure, named for French scientist Blaise Pascal (1623–1662).

SI prefixes

The SI system includes a way of making bigger and smaller units as required. It would not be convenient to express, for example, the size of a human body cell in terms of meters. It is far too small. Instead, scientists use the micrometer as a unit. *Micro-* is called a prefix, which means that it can be added to the beginning of another word. It means "millionth," so a micrometer is one-millionth of a meter. A body cell is typically between 10 and 30 micrometers across.

Modern scientific measurement is so advanced that scientists need to use some incredibly small and some incredibly large units. There are many such prefixes for SI units (see the box on page 930).

▶ *A steel tape measure is precise enough for building workers and DIY enthusiasts to measure distances accurately.*

See also: MASS AND WEIGHT

Medical imaging

Modern medical imaging technology allows physicians to look inside patients' bodies without the need for invasive surgery.

Physicians often need to look inside a patient's body to see what is causing an illness. One way of doing this is by surgery, which involves cutting open the body and looking at the bones and organs inside. However, this is a painful and dangerous process. Physicians and scientists have, therefore, developed a wide range of imaging techniques that let them look inside the body without the need for invasive surgery. They use these techniques to help diagnose (identify) a disease, see how badly a bone is broken, or help them cure an illness.

Radiology

Physicians specializing in medical imaging are called radiologists. Radiologists use a variety of techniques to make images, or scans, of the insides of a patient. The most common techniques use X-rays, but others include computerized tomography (CT), magnetic resonance imaging (MRI), positron emission tomography (PET), and ultrasound.

All of these techniques involve sending some sort of wave or beam into the body and then detecting what comes out the other side or is reflected back. The structures inside the body will absorb or reflect the wave or beam in different ways. Many internal structures, especially soft organs, may not show up well using some of these imaging techniques. With the range of technologies available to them, however, radiologists can now look at almost anything inside the body.

X-ray images

Many medical images are made using X-rays. X-rays are an invisible form of radiation. They behave like other forms of radiation, such as light

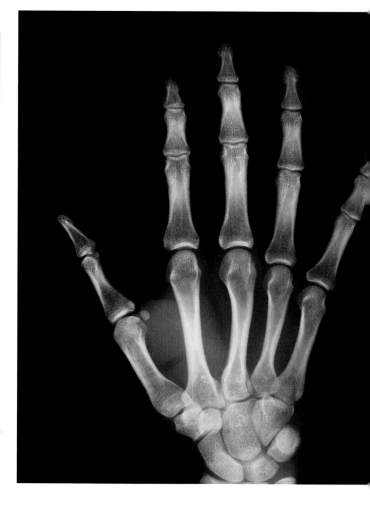

▲ *This X-ray shows the bones of the hand. The film turns black where X-rays hit it and stays white where bones block the rays.*

and heat, but they have a lot more energy. With so much energy, X-rays can pass through many solid objects, including the human body.

X-rays were discovered by German physicist Wilhelm Röntgen (1845–1923) in 1895. He named the mysterious waves he had detected X-rays because he knew very little about them. One of the first things Röntgen found out about X-rays was that they passed through the skin and revealed objects inside the body. He made the first X-ray picture in 1895 by shining the radiation through his wife's hand onto photographic film. The image clearly showed the bones in his wife's fingers as a negative image.

DID YOU KNOW?

The human body produces heat, or infrared radiation. Physicians use a machine called a thermograph to turn this heat into a thermogram. These images show areas of the body that are unusually hot or cold—a sign that something may be wrong inside. Thermography may be used to investigate arthritis (sore joints) and problems with blood circulation. Breast cancers also often show up on thermograms. Older women can be screened using thermography to check that they are not developing cancer.

Physicians began to use X-rays soon after. In 1900, surgeons with the British Army fighting in Sudan used X-rays to locate bullets inside wounded soldiers. In 1901, Röntgen received the first ever Nobel Prize for physics for his discovery.

X-ray images are formed in the same way as photographs, except that the film is sensitive to X-rays not visible light. The images reveal certain structures in the body because of the way some substances absorb X-rays, while others let them pass right through. X-rays are very good at showing bones and other hard objects, such as teeth, because these body parts do not let X-rays pass through them. However, the soft tissues around the bones, such as muscles, do let the radiation through and shine onto the X-ray sensitive film.

When X-rays hit the film, they turn it black. Parts of the film behind the soft tissue therefore turn black as the X-rays pass through the body. Parts of the film behind the bones, however, are not hit by any X-rays, so they stay white. This process makes a picture on the film that shows the areas of bone in white, with the areas of soft tissue in black. If the bone is broken, however, X-rays will penetrate the crack, and this will show up on the final image.

▶ *Physicians can image soft tissues, such as the colon (part of the intestines), by giving the patient a liquid enema containing powdered barium. The barium fills the intestine and blocks the X-rays.*

Taking an X-ray

X-ray machines are often used in hospital emergency rooms to investigate broken bones or by dentists to look at teeth under the gums. The patient must stay very still as the radiologist positions the machine to take the images. The radiologist might cover certain areas of the body with heavy flaps. These flaps prevent X-rays from entering the body and may protect delicate areas or make the picture clearer.

The machine produces X-rays for just a fraction of second. This is enough to take a picture of the patient's insides. The radiologist may have to develop the exposed film in the same way as he or she would a photograph. However, modern X-ray machines are linked to a computer, so X-ray images can be viewed immediately on a computer monitor.

Exposure to highly energetic X-rays for too long is dangerous. One X-ray picture uses only a very tiny amount of radiation, about the same amount

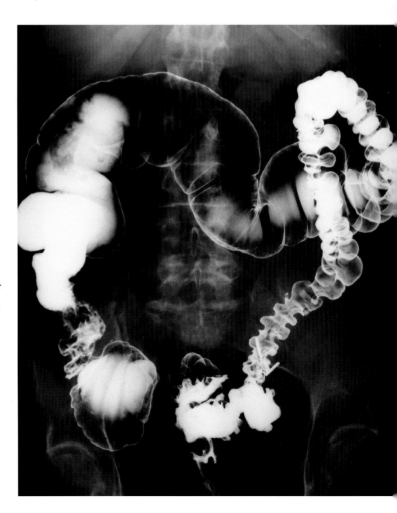

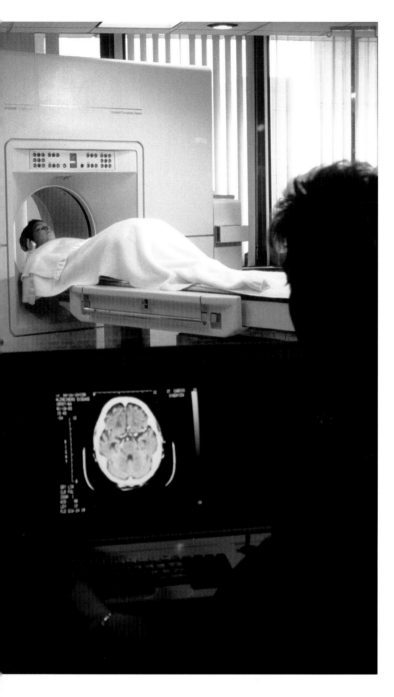

CT scans

Computerized tomography uses X-rays in a more complex way. A CT scanner takes many X-ray images from different angles, and then a computer combines the images to create an image of a slice, or cross section, through the patient's body. This cross section is viewed on a video screen in the scanner control room. Unlike simpler X-ray images, CT scans can reveal soft tissues, such as the brain, or organs such as the liver. The computer can also combine many individual CT scans to create a three-dimensional image of the inside of the body.

The patient lies on a table, and a ringed scanner, called a gantry, moves along his or her body. The area of the body to be scanned is positioned at the center of the gantry. A beam of X-rays is produced at one point on the gantry. This beam passes through the patient and is picked up by an array of detectors under the patient. The gantry rotates around the patient, and the X-ray beam is flashed through his or her body at many different angles. At each angle, the beam passes through a different amount of soft and hard tissues as it travels through the body, which may absorb more or less of the radiation. On the other side, the detectors measure how much of the beam gets through at each angle. All this information is processed by the computer to compose the final image.

Substances called contrast agents, which help certain soft organs show up on X-ray images, are often used to make CT scans of certain body parts clearer. The contrast agents are harmless substances that absorb X-rays. A liquid containing barium is used to highlight the stomach and intestines. It is swallowed by a patient before the scan. Iodine compounds may be injected into the blood if the radiologist needs to look at the blood supply to a patient's brain or another vital organ.

▲ *In this photograph, a radiologist studies on screen a slice across the brain, which has been revealed by a CT scan. Through the window, you can see the patient's head in the gantry, where the X-rays probe the brain to produce the image.*

someone receives making a short flight in a passenger jet. Having a few X-rays taken now and then is not thought to be dangerous. But radiologists are taking X-rays all the time, and so they usually stand behind a protective screen while the procedure is being done.

Seeing with magnets

Magnetic resonance imaging uses magnets and radio waves to form images of soft body tissues. Since it does not use dangerous radiation like many other forms of medical imaging, MRI scans are generally used to study the most delicate parts of

DID YOU KNOW?

Angiography is one form of X-ray imaging that is used to look inside blood vessels. Physicians use angiography to check if important blood vessels, such as those supplying the heart or brain, are blocked or about to burst. First, a contrast agent is injected into the blood vessel. The contrast agent is generally an iodine-containing substance that helps the blood vessel show up when the X-ray image is taken.

the body, such as the brain and heart, and to look inside the eyes and ears. MRI scans can also make images of soft tissues, such as the cartilage in joints, that do not show up well in X-rays or CT scans. MRI scanners are very expensive, so only the largest hospitals use them.

An MRI scanner has a hollow tube at its center where the patient lies. The tube is surrounded by a powerful electromagnet (a magnet that can be turned on and off), a source of radio waves, and many radio-wave receivers. When the electromagnet is switched on, many of the atoms inside the patient's body align themselves with the strong magnetic field. The patient is not aware of this painless change. Since they use a strong electromagnet, however, MRI scanners cannot be used by people who have metal implants inside the body. These implants may be replacement joints or electrical devices, called pacemakers, that keep hearts beating regularly.

Radio waves, including the ones that carry radio and television signals, are a type of radiation. They behave in exactly the same way as light, heat, and X-rays. But radio waves carry only a small amount of energy. While the electromagnet is on, the MRI scanner produces a large pulse of radio waves, which passes through the patient's body.

▶ *This MRI scan shows all the structures inside the head, including the large nasal cavity behind the nose and all the convoluted folds of the cerebrum—the part of the brain used for conscious mental processes.*

When a radio wave hits one of the many hydrogen atoms aligned with the magnetic field, the radio wave transfers some of its energy to the hydrogen atom. This extra energy makes the hydrogen atom wobble out of position. When the radio pulse is switched off, the wobbling hydrogen atom returns to it original position, lining up with the scanner's magnetic field again. As it moves back into this position, the hydrogen atom releases its own radio waves, which pass out of the body and are detected by the receivers in the scanner. The size and length of these radio waves depends on which type of body tissue they have passed through. A computer analyzes the radio waves coming from the patient and uses them to produce a very detailed image of the interior of the body. Radiologists view this image on a video screen.

Live action

Most types of medical imaging produce still images. Although these may be very clear, they cannot show how a body part is working. Positron emission tomography can produce moving images that show the activity of a living organ.

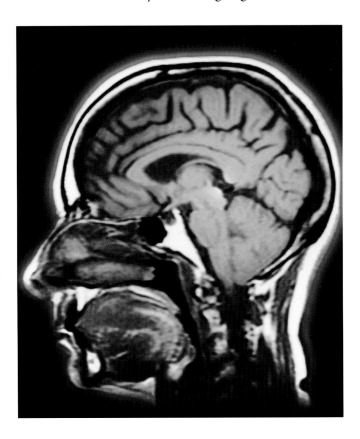

PET scanners are generally used to study brain activity. Before the brain is scanned, the patient is injected with radioactive sugar. The amount of radiation produced by this sugar is harmless.

The sugar molecules travel to the brain, where they can be detected by the scanner. The radioactive component of the sugar molecules emit positrons. Positrons are tiny particles that are identical to electrons, except that they have a positive charge. (Electrons are always negatively charged.) The positrons collide with electrons in the brain and produce gamma rays. Gamma rays are another form of high-energy radiation, similar to X-rays. The PET scanner is set up to detect the gamma rays radiating out of the patient's head in all directions and to map their pattern.

▼ *This PET scan reveals the damage done by the narcotic cocaine. The top three scans show activity in a normal brain in red and yellow. The bottom three show the reduced brain activity associated with cocaine dependency.*

DID YOU KNOW?

An endoscope is a long, thin device that uses fiber optics to look inside the body. Endoscopes can be inserted into a patient's body through the mouth or another body cavity or through a small incision. Some endoscopes carry a tiny camcorder inside the body that relays moving images to the surgeon performing the procedure.

Endoscopes are used to investigate the lining of the stomach or intestines or to look for problems elsewhere that cannot be seen using other imaging techniques. As well as taking pictures, endoscopes can also be fitted with surgical instruments, such as scalpels or tiny lasers that are used to cut away diseased tissues. This type of surgery is called keyhole surgery, because only a small incision is made in the patient.

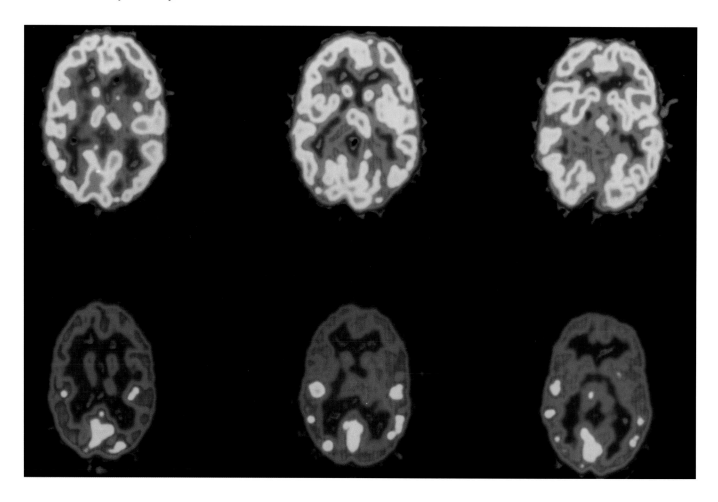

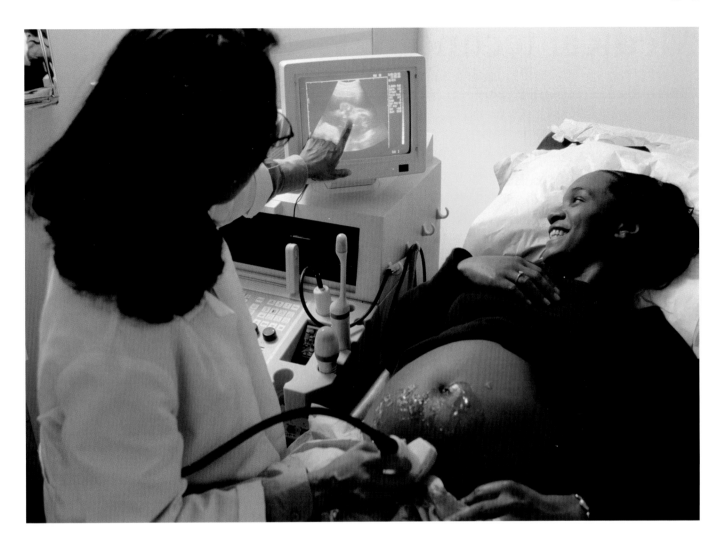

▲ *Most expectant mothers now have an ultrasound scan done to show how their unborn babies are developing in the uterus. Problems can then be diagnosed and possibly treated.*

The area of the brain that is most active has the most sugars in it and produces more gamma rays than less active parts. Physicians use PET scanners to investigate which parts of a person's brain are used to control different activities, such as moving and thinking. Using PET techniques, physicians have discovered that different parts of the brain are linked to emotions. PET scans also help reveal brain damage caused by a stroke or disease.

Ultrasound

Ultrasound is sound that is too high-pitched for people to hear. Sound waves are not the same as radiation such as visible light or X-rays. Instead, sound waves are the atoms in a substance squeezing together and then spreading out. The ears detect compressions and expansions in the air, and the brain turns them into sounds.

An imaging device called an ultrasound transducer turns electricity into ultrasonic waves. When the transducer is pressed against the body, the waves travel into the body. When they hit a solid surface, some of the ultrasound waves bounce back in the same way as an echo. The transducer picks up this echo, and a computer processes the data into an image on a monitor. Ultrasound is harmless, so it is used to take pictures of delicate parts of the body, such as the genitals. Ultrasound is also used to look at a fetus growing inside the uterus. Many parents are now given a DVD or photograph of the ultrasound scan of their unborn baby.

See *also:* MEDICAL TECHNOLOGY • X-RAY

Medical technology

Medical technology is the application of science to medicine. Technology plays a vital role in all aspects of modern medicine, from the diagnosis and prevention of disease to drug treatment, radiotherapy, and surgery.

Greek scholar Hippocrates (c. 460–c. 377 BCE) is the founding father of modern medicine. Unlike his contemporaries, who thought that disease was a punishment from the gods or caused by evil spirits, Hippocrates suggested a rational explanation for disease based on natural causes. He stressed the importance of the body's own healing process and prescribed cleanliness, a good diet, and plenty of rest as a cure for all ailments.

A scientific basis for medicine came in 1543, when Belgian anatomist Andreas Vesalius (1514–1564) published the first detailed account of human anatomy. Then, in 1628, English physician and anatomist William Harvey (1578–1657) described the circulation of blood in the body. Another major development was the invention of the microscope, which would later help French chemist Louis Pasteur (1822–1895) develop his germ theory of disease. By this time, English physician Edward Jenner (1749–1823) had also developed the concept of vaccination.

Medical technology progressed rapidly in the nineteenth century. Following Pasteur's discovery of germ theory, English physician Joseph Lister (1827–1912) stressed the need for antisepsis during surgery. Patients also benefited from the development of anesthesics such as ether, nitrous oxide, and chloroform. A number of diagnostic tools were also developed, such as the stethoscope, X-rays, and the sphygmomanometer. Computer technology and genetics have revolutionized modern medicine, but they have also raised some serious ethical dilemmas.

Medical imaging equipment

Medical imaging equipment enables physicians to look inside the human body without the need for invasive exploratory surgery. Medical imaging is an essential part of medical diagnostics. Different scanning machines record images of different parts of the human body.

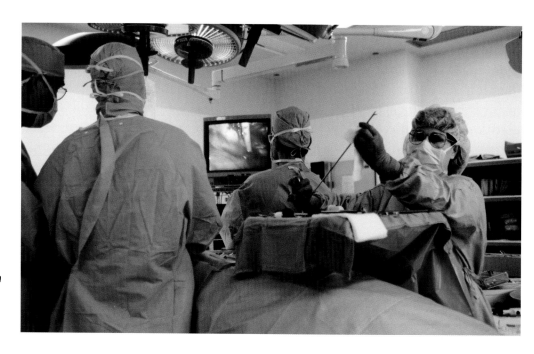

▶ *A surgeon inserts an endoscope through a small incision in the patient's body. A camera mounted at the end of the endoscope lets the surgical team view the inside of the body on a monitor. The surgeons can then operate using surgical instruments mounted alongside the camera. Such minimal access surgery (MAS) is much safer and less stressful on the body than conventional surgery.*

German physicist Wilhelm Röntgen (1845–1923) discovered X-rays in 1895. The invention of the X-ray machine followed soon after. X-rays are good at taking images of dense body parts, such as bone. As X-rays pass through the body, dense body parts absorb X-rays and cast a white shadow on the developed film. In this way, X-rays can reveal the extent of damage to a broken bones, blockages in blood vessels, and foreign objects inside the body.

Physicians recognize that there is a small risk associated with exposure to X-rays. Ultrasound scanning is a much safer alternative. Ultrasonic equipment uses sound waves that are too high-pitched for people to hear. Ultrasonic waves can be used to create images of the inside of the heart, uterus, or other delicate organs. The waves are transmitted through the body to the organ being studied. Some of the waves are reflected back by the organ, and these "echoes" form an image on a computer monitor. Brain damage and disease can often be detected by means of ultrasonic images.

More recent developments in medical imaging include computerized tomography (CT), magnetic resonance imaging (MRI), and positron emission tomography (PET). CT uses X-rays to construct three-dimensional (3D) images of the interior of

A radiologist studies CT scans of the brain to detect the extent of brain damage following a stroke. The technique was developed in the 1960s by English electrical engineer Godfrey Hounsfield (1919–2004).

the body. MRI uses powerful magnetic fields and radio waves to construct highly detailed 3D images of tissue structure. PET scans reveal body functions, such as brain metabolism, by measuring the body's uptake of radioactive substances.

MEDICAL MONITORING EQUIPMENT

Devices that measure different aspects of physiology (body function) are called medical monitors. These devices record functions such as blood pressure, heart rate, breathing rate, and electrical activity of the brain. Medical monitors play a vital role in diagnosis as well as in monitoring patients under the influence of an anesthetic during surgery.

Stethoscope

One of the first things that physicians do when examining a patient is to listen to his or her chest with a stethoscope. This device helps physicians hear the sounds made by the heart and lungs, which indicates whether they are healthy or not.

The idea of the stethoscope came to French physician René Théophile Laënnec (1781–1826) in 1816. His instrument was a thin wooden tube, about 1 foot (30 centimeters) long, with a bell-shaped opening at one end. The modern stethoscope, developed by U.S. physician George P. Cammann (1804–1863), has two earpieces at the tips of two metal tubes, which are connected to a metal chest piece by flexible rubber tubes. The chest piece has two parts. One is a shallow metal bell, which picks up many different sounds in a wide range, but especially low-pitched tones. Around this bell is a plastic diaphragm (thin membrane), which picks up high-pitched tones made as the heart beats and the lungs breathe.

Blood pressure monitors

It is often important to measure a patient's blood pressure to check whether the heart is circulating the blood properly. A sphygmomanometer is used

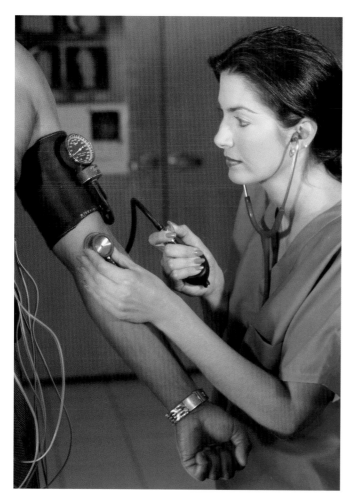

▲ A physician measures the blood pressure of her patient using a sphygmomanometer and stethoscope.

to measure blood pressure as it runs through the arteries. This device was invented in 1896 by Italian physician Scipione Riva-Rocci (1863–1937).

The sphygmomanometer consists of a tube that connects at one end to a long inflatable bag, called the cuff, which is wrapped around the upper arm. The person taking the blood pressure inflates the cuff using a small hand pump until the pressure exerted on the forearm is just enough to stop blood from flowing through a major artery in the forearm, just below the cuff. This point is detected by listening through a stethoscope to the blood pulsating. The difference in levels in the sphygmo-manometer, shown on a scale, is a measure of the blood pressure. The sphygmomanometer records two pressures. The higher (systolic) pressure occurs when the heart contracts; the lower (diastolic) pressure occurs when the heart rests between beats.

Electrocardiograph

The electrocardiograph (ECG) records electrical signals from the heart muscles as they pump blood around the body. The ECG was invented by Dutch physiologist Willem Einthoven (1860–1927) in 1903. The signals from the heart are picked up by electrodes fixed to the person's chest. Readings may also be taken from the arms and legs. The ECG amplifies (strengthens) the signals so that they can be displayed on a screen or cause a pattern of the heart waves to be traced on paper. The ECG equipment also contains a device that measures the rate at which the heart beats.

ECG machines are used to check the condition of a person's heart during a medical examination. They are also used to monitor the activity of the heart after an accident, surgery, heart attack, or other serious illness. In such cases, the machine keeps a continuous check on the heart rate.

Alarm circuits automatically warn the medical staff if the heartbeat becomes abnormal. At the same time, a sensitive lever starts to draw out the wave patterns. The recordings help physicians decide what is wrong with the patient's heart.

Fetal heart-rate monitor

Pregnancy and the birthing process are stressful times for both mother and baby. It is common for physicians to monitor the unborn baby's heart rate to see if the baby is being deprived of oxygen—a condition called fetal distress. Fetal distress often occurs during labor, when less oxygen reaches the baby from the placenta. Occasionally, the umbilical cord becomes wrapped around the baby's neck. For a healthy unborn baby, the heart rate is usually between 120 and 160 beats per minute. With fetal distress, the baby's heart rate does not fluctuate but falls to a fixed lower level.

The monitor sends ultrasound waves through the mother's body to the baby. The baby's beating heart reflects the ultrasound waves, and the echoes are detected by the monitor and translated into an image on a screen. Alternatively, the heart rate can be measured by means of an electrode, which is placed on the baby's head through the vagina. The electrode detects electrical signals from the unborn baby's heart, and these signals are translated into the heart rate measurement.

If the unborn baby is severely distressed, he or she may be delivered by cesarean section. If labor is more advanced, the baby may be delivered using forceps or a device called a vacuum extractor.

Electroencephalograph

In 1929, German biologist and psychiatrist Hans Berger (1873–1941) introduced the electro-encephalograph (EEG), which is a machine used to study brain waves. These are small electrical signals that the brain generates all the time. Brain waves are detected by means of metal contacts, called electrodes, usually fixed to the patient's scalp. During brain surgery, however, the electrodes may be placed directly on the surface of the brain. The signals detected by the electrodes are amplified in the EEG machine and displayed on a monitor. The pattern of these brain waves depends on the activity of the brain. In turn, this depends on the health of the patient and what he or she is doing at the time. Healthy brains give out waves with similar patterns. An unusual wave pattern usually means that the brain is diseased or has been injured. The EEG machine is also used for research into the functions of the various parts of the brain.

Medical sensors

Many hospitals now have automated electronic equipment, called medical sensors, that measure and record physiological functions such as blood pressure. These devices contain an electrical component, called a transducer, which translates the physical measurement into an electrical signal. The results are then displayed as a digital readout on a monitor. Medical sensors often sound an alarm when the measurement exceeds or falls below a normal level.

Biosensors are the medical sensors of the future. They are used to record the levels of body chemicals, such as blood glucose levels. A biosensor uses an enzyme (biological catalyst) to react with the chemical being measured, resulting in a product that may be detected by a probe, for example, through a change in color or acidity. Then, in the same way as a medical monitor, a transducer in the probe converts the measurement into a digital readout on a monitor.

Laboratory tests

Lab tests are a vital tool for medical monitoring and diagnosis. Microscopes are routinely used in pathology labs to study tissue samples. Biosensors are increasingly being used to perform blood tests to measure the levels of enzymes and chemicals such as glucose. The results of laboratory tests indicate the performance of organs such as the kidneys and the liver.

DID YOU KNOW?

When a person's heart stops beating, immediate action should be taken to revive the patient. This can be done using a device called a defibrillator. A defibrillator works by giving the heart a powerful electric shock using two electrodes that are held on the chest, over the heart. The shock makes all the patient's chest muscles contract (tighten up), including those of the heart. This will often cause the heart to start beating in a regular manner again. Sometimes it is necessary to apply a shock several times.

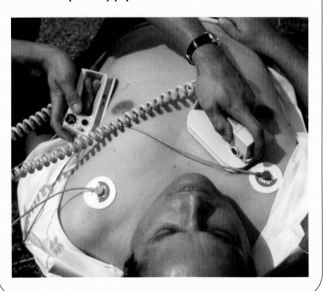

▲ *Alexander Fleming, a Scottish physician, accidentally discovered the antibiotic penicillin in 1928.*

DID YOU KNOW?

Ventilators are machines that assume the respiratory function of patients who cannot breathe for themselves. They consist of an electric pump connected to an air supply. The pump forces air along a tube that passes into the lungs through the patient's airways. Before the air enters the lungs, it passes through a humidifier, which adds sterile water vapor to keep the lungs moist. The natural elasticity of the lungs and rib cage helps expel the exhaled air into an outlet tube. A one-way valve fitted to the outlet tube prevents the exhaled air from being rebreathed.

Intensive care unit

Patients with serious illnesses or injuries are treated in intensive care units (ICUs). ICUs are equipped with a wide range of electronic devices for checking the patients' health. Each set of equipment is called a patient monitoring system. Several separate machines may be used in such a system, or they may be combined to form a single unit.

The ventilator, or life-support machine, is an essential part of the ICU. It provides enough oxygen to meet the demand of the patient's main organs, such as the brain, kidneys, and heart. Medical staff also use routine equipment, such as the sphygmomanometer and heart-rate monitor, to check on the condition of patients in ICU.

MODERN MEDICAL TREATMENT

Medical professionals will decide on the best course of treatment based on an accurate diagnosis and the perceived benefits of the treatment plan. Medical treatment may include drug therapy, radiation therapy, surgery, or a combination of the three. Medical staff may also recommend the use of devices that support certain body functions, such as kidney dialysis machines and pacemakers.

Drug therapy

Drug therapy, or chemotherapy, is as old as medicine itself. Ancient Chinese medicine relied (and still relies) on a wide range of herbal remedies. Elsewhere, early drugs often contained ingredients such as feces, sweat, and urine. During the Renaissance, Swiss physician Paracelsus (1493–1541) popularized the use of compounds such as arsenic, mercury, and the opium derivative laudanum. Herbal remedies were also widely used in Europe at this time.

Antimicrobial drugs first appeared at the beginning of the twentieth century. In 1909, German physician Paul Ehrlich (1854–1915) discovered salvarsan—a compound of arsenic successfully used to treat the bacterial disease syphilis. Then, in 1928, Scottish physician Alexander Fleming (1881–1955) discovered penicillin. The golden age of drug development began in the 1930s, when German scientist Gerhard Domagk (1895–1964) discovered that some of the deadliest bacterial diseases could be successfully treated with a sulfa-containing dye.

Today, the drug industry is highly sophisticated. Computers control all stages of drug development, from investigating a drug's chemical structure to the synthesis process. Perhaps the greatest development in modern drug therapy is in the delivery technology. Drugs can now be taken in the form of eyedrops, lotions, and inhaled sprays.

Radiation medicine

For more than one hundred years, physicians have been using X-rays to look inside the human body. Today, radiation is also used to treat diseases such as cancer. Cancer occurs when body cells divide uncontrollably, leading to the formation of growths called tumors. Radiation can be used to destroy cancer cells, but it also destroys healthy body cells. Therefore too much radiation in the wrong place can damage the body or even cause new cancers. The aim of radiotherapy, as the technique is called, is to give the right dose of radiation to the unhealthy cells that need to be destroyed. Healthy tissues must be avoided. Body scans now make it possible to see exactly where a tumor is, so that radiologists can target the X-ray beam onto it. Further scans then show how successful the treatment has been.

Sometimes, rather than use a beam of X-rays, the source of radiation may be surgically implanted inside the body. Radioactive iridium wires, implanted into the body, can deliver a continuous small dose of radiation exactly where it is needed.

Another technique that combines body scans with X-ray treatment is interventional radiology, in which radiologists perform procedures directly inside the body. They use CT scans or ultrasound to guide them when inserting fine wires or instruments into the body. Surgery that once required large, invasive incisions now leaves only a small puncture mark. Patients often do not need a general anesthetic and can be operated on and return home on the same day.

One example of this is balloon angioplasty. This is a method of opening up blocked or narrowed blood vessels. First, the surgeon feeds a guide wire into the blood vessel, followed by a tube called a catheter. This has a type of balloon on its end. This can be inflated to remove the blockage, which is usually a fatty material called atheroma. When the main artery to a leg is affected, balloon angioplasty can restore a good blood supply to the whole limb.

Surgery

Most surgery is now routine, thanks to the use of anesthetics, medical monitoring equipment, and technology such as the heart-lung machine. General anesthetics render the patient unconscious and unable to feel pain. Muscle relaxants make it much easier for surgeons to operate on body parts. Medical staff use monitoring equipment to check on the health of the patient during surgery. This is especially important for patients under general anesthesia. A heart-lung machine assumes the circulatory and respiratory roles of the heart and lungs during open-heart surgery.

Organ transplants were first performed in the 1950s, with limited success. Today, this type of surgery is almost routine in the developed world.

▲ *During the Middle Ages, surgeons used an array of gruesome instruments, such as hooks and saws, to perform operations. Without the use of anesthetics or an understanding of infection, surgery inevitably led to the deaths of most patients.*

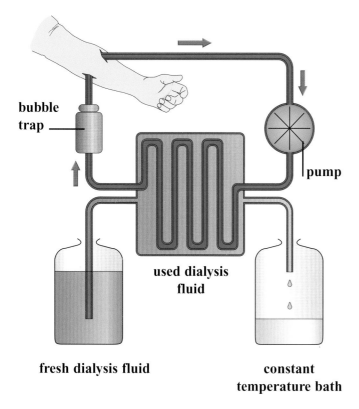

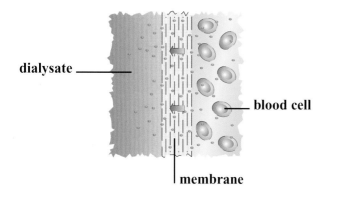

▲ *Dialysis was first used in 1943 to treat human patients suffering from acute or chronic kidney failure.*

However, organ transplantation is limited by the supply of organs and the probability of organ rejection. To overcome these limitations, medical researchers are experimenting with tissue culture (growth) using cells from the patient's body. For example, scientists have grown bladders in the laboratory by culturing bladder tissue around a mold. Complex organs, such as the heart, pose a greater problem. In this case, the most probable application of the new technology is to replace damaged areas of the heart with cultured tissue.

Artificial body parts and aids

In many cases, artificial devices can be used to support or assume the role of damaged body parts. Simple aids include eyeglasses and hearing aids, but the use of more complex devices, such as pacemakers to regulate the heart, is almost routine.

Machines can even assume the function of organs such as the liver and kidneys. The first kidney dialysis machine was developed by Dutch physician Willem Kolff (1911–) in 1943. Effectively, the kidney dialysis machine is an artificial kidney. It takes over the kidneys' role of removing toxic substances and excess water from the body. It does this by filtering the blood through a semipermeable membrane. The impurities in the patient's blood diffuse through it into the dialysis fluid. Dialysis has to be done for periods of around four hours, three times a week. Many patients are on dialysis for years—or even for the rest of their lives. Nevertheless, dialysis is still regarded as a temporary measure. The only permanent solution is a kidney transplant.

A new technique, called continual ambulatory peritoneal dialysis, avoids the need for the dialysis machine. The patient first needs to have a tube inserted into his or her abdomen. To begin the treatment, the patient runs about 4 pints (2 liters) of dialysis fluid through the tube into the peritoneal cavity, which surrounds the abdomen. The impurities now diffuse straight from the membranes in the abdomen into the dialysis fluid. A few hours later, the patient drains the fluid off and throws it away. The patient can continue his or her daily routine while the process is going on inside the body.

COMPUTERS IN MEDICINE

Computers are widely used in hospitals to store and update records of patients, equipment, drugs, and other supplies. Increasing use is being made of computers to diagnose diseases from given signs and symptoms so that the correct treatment can be given at once. Computer technology has also empowered the patient. People with a wide range of medical conditions can join support groups on the

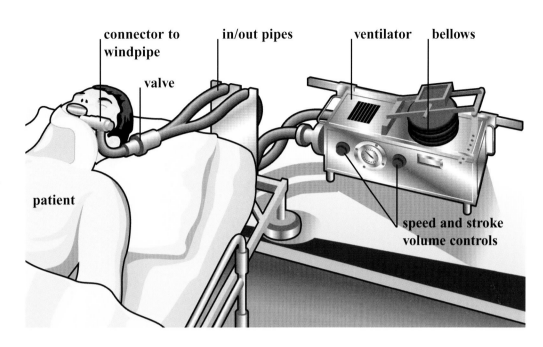

▶ This illustration shows an artificial ventilation machine, which is used when a person can no longer breathe naturally. In this process, air is transmitted to the patient's lungs through a tube inserted into the windpipe. After the lungs are filled, the air is expelled by the natural elasticity of the lungs.

connector to windpipe

in/out pipes

ventilator

bellows

valve

patient

speed and stroke volume controls

Internet and browse medical Web sites aimed at the nonspecialist. In this way, patients can learn more about their medical conditions and find out about the latest treatments.

MEDICAL ETHICS

The use of some medical technologies raises some ethical dilemmas. For example, how long should life-support machines be used to keep alive premature babies and the victims of traffic accidents? Who decides when the machine should be switched off? The use of genetics in medicine is also controversial. It is now possible to screen an unborn baby for genetic diseases, but many people are concerned about how this information will be used. For example, is it acceptable to abort a fetus with genetic abnormalities? Stem cell research is another sensitive medical issue. Stem cells are cells that can grow into any one of about 300 different types of human body cells. Stem cells are found in both adults and human embryos, but most cells for research are taken from embryos that have been discarded during fertility treatments. Ethical groups and antiabortionists oppose the use of stem cells, because the embryos are destroyed after the stem cells have been isolated. The creation of embryos solely for the production of stem cells is even more controversial.

DID YOU KNOW?

Lasers can be used to perform extremely delicate surgery. They are used in two ways—to cut like a scalpel or to burn away areas of diseased or damaged tissue. The laser beam is totally aseptic (germ-free). With the beam set at a very narrow width, it can be used as a scalpel. It has the advantage of sealing up minor blood vessels and thus reduces blood loss and shock for the patient. Indeed, the laser had been dubbed the "bloodless scalpel" by many surgeons. Using a laser in this way demands great skill and precision. With the laser beam at a width of a few inches, it can destroy or remove areas of unhealthy tissue at a far greater speed than any other surgical tool. Recovery of the patient is also much quicker than when other techniques are used.

See also: ANESTHETIC • ANTIBIOTIC • ANTISEPTIC • CLONING • DENTISTRY • GENETIC ENGINEERING • GENETICS • MEDICAL IMAGING • RADIOTHERAPY • SURGERY • X-RAY

Mendel, Gregor

Austrian monk Gregor Mendel combined his love of mathematics and biology in botanical research, mainly in the breeding of pea plants. Through his research, Mendel found one of the missing pieces in Charles Darwin's theory of evolution and laid the foundations for modern genetics.

Gregor Johann Mendel was born on July 22, 1822, in the Austrian town of Heinzendorf (now Hyncice in the Czech Republic). After studying at the Philosophical Institute in Olmütz (now Olomouc), Mendel moved to the Augustinian monastery in Brünn, Moravia (now Brno in the Czech Republic), in 1843. During his time at the monastery, Mendel became interested in science and mathematics and started to teach at a local secondary school. Mendel was ordained as a priest in 1847 and, seeing his vocation as a teacher, took an examination to get a teaching license. Mendel failed the exam, however, so he enrolled at Vienna University to study science and mathematics. In 1853, Mendel returned to the monastery in Brünn to continue his religious studies and teach at the local Technical High School.

Experiments with pea plants

In 1856, Mendel embarked on a series of experiments that would lead to the discovery of the basic laws of genetics. In the gardens of the monastery, Mendel cultivated more than 30,000 specimens of pea plants (*Pisum* sp.), carrying out his research over eight years. Mendel was meticulous in his preparation. With the help of two full-time assistants, Mendel self-pollinated and wrapped each plant to prevent cross-fertilization and pollination by insects. He then collected seeds from these plants and examined the plants produced by the seeds. In one experiment, he

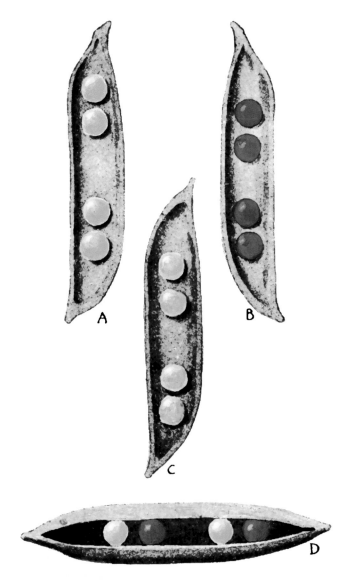

▲ *In his experiments into heredity, Mendel cultivated peas that produced yellow (A) and green (B) peas. This produced a generation in which the peas are yellow (C). By breeding generation C peas together, Mendel found that the next generation (D) had a mixture of pea colors. Experiments such as this one led to the discovery of the basic laws of heredity.*

crossed a dwarf pea plant with a tall pea plant and planted the seeds. He found that all the resulting pea plants were tall—no dwarfs grew. When he self-pollinated the new tall pea plants, however, the next generation was a mixture of three tall plants to every one dwarf plant. In other words, the factor determining tallness was more dominant (stronger) than the factor determining dwarfness, while the dwarfness was recessive (weaker) and only came out in the next generation. Mendel called the "factors" *Elemente* (now called genes).

Mendel studied many other characteristics of the pea plants, such as flower color and position, seed shape, and pea color. He then confirmed his findings with similar experiments on corn, flowers, and other plants. As with the height experiment, Mendel found that a simple set of rules could predict the inheritance of the traits.

Mendel thus formulated the fundamental laws of heredity. Most naturalists at the time assumed that crossbreeding would produce an amalgamation of characteristics. This theory predicted that crossing a tall plant with a dwarf plant would produce a pea plant intermediate in size between the two parent plants. In his theory of evolution, English naturalist Charles Darwin (1809–1882) recognized that occasionally new traits or combinations of them (mutations) would appear and be passed onto their offspring, but he had no explanation for why they occurred. It was not until the mid-twentieth century with the discovery of the genetic code in deoxyribonucleic acid (DNA) that both Mendel's laws and Darwin's variations could be explained.

Return to religion

In 1865, Mendel reported the results of his experiments to the Natural History Society of Brünn. A year later, he published a scientific paper. However, the importance of the work was not fully recognized during Mendel's lifetime. Even respected Swiss botanist Karl-Wilhelm von Nägeli (1817–1891) dismissed the paper.

> **DID YOU KNOW?**
>
> Although it is widely reported that Mendel's work was ignored during his life time, the *Encyclopedia Brittanica* published a favorable account of his studies. Mendel was also frequently credited in the work of German botanist Wilhelm Olbers Focke (1834–1922).

▲ *Austrian monk Gregor Mendel, pictured here at the age of 40, laid the foundations for modern genetics.*

Mendel abandoned his research in 1868, perhaps frustrated with the lack of recognition for his work, but most likely because his promotion to abbot of the monastery at Brünn left him little time for science. He continued as abbot at Brünn until his death on January 6, 1884.

Mendel rediscovered

Mendel finally achieved recognition in 1900, when three scientists—Dutch botanist Hugo de Vries (1848–1935), German botanist Carl Correns (1864–1933), and Austrian botanist Erich von Seysenegg (1871–1962)—rediscovered his work. When de Vries tried to publish Mendel's work as his own, his two rivals made it clear to the scientific community that credit really belonged to Mendel.

See also: EVOLUTION • GENETICS

Mendeleyev, Dmitry

In 1871, Russian chemist Dmitry Mendeleyev published a system of arranging the chemical elements in groups on a chart called the periodic table of the elements.

Dmitry Ivanovich Mendeleyev was born in Tobolsk in the west of Siberia on February 7, 1834. He was born to a well-educated, prosperous, and very large family. (Dmitry was the youngest of between 14 and 17 children.) When he was a child, Mendeleyev's father became blind and was forced to resign his position as headmaster of the local high school. To support the family, Mendeleyev's mother went to work at a glass factory in a nearby town. Mendeleyev attended a school in Tobolsk and was a promising student, preferring mathematics and physics to the compulsory classics.

Unfortunately, the Mendeleyevs were plagued by bad luck. When Mendeleyev was 13, his father died. A year later, the glass factory where his mother worked burned down. When Mendeleyev left school at the age of 16, his mother decided to take him to university to finish his education. With little money, mother and son were forced to hitchhike the 4,000 miles (6,400 kilometers) to Moscow. When they got to there, Mendeleyev was turned away. Moscow University did not admit people from the "backward province" of Siberia. Undeterred, they continued to St. Petersburg but, again, both the university and medical school refused to admit the young Mendeleyev.

Eventually, Mendeleyev's mother found him a place at the St. Petersburg Institute of Pedagogy, where he trained to become a teacher. There he was educated by Russian chemist Alexander Woskressensky (1809–1890), whose research in inorganic chemistry fascinated Mendeleyev's bright mind. Mendeleyev also attended classes in mineralogy and zoology.

Career in chemistry

Mendeleyev received his teaching diploma in 1855, receiving a gold medal for his academic excellence. He then worked as a science teacher, first in Simferopol and then at Odessa. In 1856, Mendeleyev returned to St. Petersburg to take a master's degree in chemistry. A year later, he was given a junior post at St. Petersburg University.

In 1859, the Russian government awarded Mendeleyev a grant that enabled the young chemist to travel in Europe, where he ended up at Heidelberg University in Germany. There, he met Italian chemist Stanislao Cannizzaro (1826–1910), who spurred Mendeleyev's interest in the physical

▲ *Russian chemist Dmitry Mendeleyev looked upon his scientific work as a duty to his country.*

▶ *Mendeleyev is pictured here at work in his study at St. Petersburg University.*

and chemical properties of groups of elements. On his return to St. Petersburg in 1861, Mendeleyev was awarded a doctorate and accepted a position as professor of chemistry at the Technological Institute of St. Petersburg. Three years later, he moved to St. Petersburg University as professor of general chemistry. Part of his duties included lecturing undergraduate students in inorganic chemistry. Mendeleyev could not find a suitable textbook to recommend to his students, so he produced his own. The resulting *Principles of Chemistry* (1870) contained the work that would bring Mendeleyev fame throughout Europe.

The periodic law

In the mid-1800s, several systems for classifying the elements were suggested. All arranged the elements according to their atomic weight. Mendeleyev followed this accepted formula and compiled a series of cards, listing the chemical properties and atomic weights of all the 63 elements then known. As he arranged the cards, Mendeleyev noticed that some of the properties of the elements were repeated at regular intervals. Elements with similar chemical properties lined up in vertical columns, now called groups, and horizontal rows, now called periods. The arrangement conveniently formed a table, now known as the periodic table of elements. Mendeleyev's system also predicted that there were new elements still to be found. This was confirmed by the discovery of gallium in 1875, scandium in 1879, and germanium in 1886. Each new element took its position on the table, and Mendeleyev's concept of periodicity finally gained acceptance.

DID YOU KNOW?

In 1955, scientists at the University of California produced a new artificial element—atomic number 101—which was named mendelevium in honor of Mendeleyev's contribution to chemistry.

Mendeleyev's periodic table is all the more remarkable given that he had no idea about the structure of atoms. It is now known that elements in the same group have the same number of electrons in the outer electron shell, and it is this property that defines the chemistry of an element. The modern periodic table arranges elements by atomic number, which usually, but not always, matches the order of increasing atomic weight. Modern printed versions of the periodic table include atomic mass data along with much other useful information about the elements.

Other achievements

As well as his important contribution to chemistry, Mendeleyev tackled problems in fields as diverse as aeronautics, agriculture, astronomy, and meteorology. He retired from academic life in 1890 following a dispute with St. Petersburg University. Three years later, Mendeleyev accepted a position as the director of the Bureau of Weights and Measures. He died in St. Petersburg in 1907, a few days before his 73rd birthday.

See also: CHEMISTRY • PERIODIC TABLE

Mercury, metal

In the Middle Ages, scientists did not consider mercury to be a metal. Instead, it was thought to be a basic element of all metals. Only in the eighteenth century, when it was proved that mercury could be frozen solid, was it accepted as a true metal.

Mercury is a silver-colored metal and the only metallic element that is liquid at standard room temperature and pressure. It is commonly called quicksilver. The chemical symbol for mercury is Hg, short for *hydrargyrum*, which comes from the Greek words *hydro*, meaning "water", and *argyros*, meaning "silver". Another key property of mercury is that it is very dense, being about 13½ times heavier than water, with an atomic weight of 200.59. Mercury is also highly toxic.

Mercury is sometimes found above ground among metallic ores. Ores are rocks containing high proportions of metals combined with other elements in the form of compounds. Since mercury ore can be found above ground, the metal was probably known to prehistoric people. The extraction of mercury from the ore cinnabar was first mentioned by Greek scholar Aristotle (384–322 BCE) in the fourth century BCE. Cinnabar has long been used as a bright red pigment called vermillion. The ancient Chinese, Egyptians, and Indians also knew of mercury. The metal's great density and its unique, liquid quality attracted medieval alchemists. Alchemists were early scientists who attempted to turn base metals into gold.

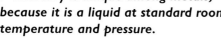

▶ *Mercury is unique among metals, because it is a liquid at standard room temperature and pressure.*

Extracting the metal

About 50 percent of the world's supply of mercury comes from Italy and Spain. Other important mercury producers are Germany, Slovenia, and the United States (in particular, the states of Texas and California). Mercury is usually taken from cinnabar by cooking the ore in a furnace. At 675°F (357°C) the mercury boils and separates from the ore by turning into vapor (vapor is the gaseous state of a substance that is usually a liquid or solid). The mercury vapor then passes out of the furnace through pipes, condenses, and collects as a liquid. Since the boiling point of mercury is relatively low, it is easy to produce very pure mercury.

In modern mercury production, all of the cinnabar is ground into a fine powder and then roasted in one of two types of furnace. In one type of furnace, the ore is moved from one hearth to successively lower ones by turning rakes. The other type of furnace is a long revolving cylinder, set at a slightly inclined angle. By the time the ore has reached the bottom of the furnace, most of the mercury has been extracted.

Uses of mercury

Perhaps the most familiar use of mercury is in the thermometer, which measures temperature. This device was invented in 1714 by German physicist Daniel Gabriel Fahrenheit (1686–1736). Thermometers make use of mercury's uniform expansion under heat and the great range between its melting and boiling points.

Another common use is in the barometer, invented by Italian physicist Evangelista Torricelli (1608–1647) in 1643. The barometer measures atmospheric pressure. More recently, mercury has been used in another measuring instrument, the manometer, which measures a person's blood pressure. Both of these devices use mercury because it is the most dense liquid available. As a result, smaller amounts of mercury can be used to measure different pressures than would be possible with other liquids, such as water, and so the equipment can be made smaller.

Mercury also has important electrical uses. One very common use is in fluorescent lighting tubes, in which the brilliant blue-white light is produced by an electrical charge being passed through mercury vapor. Since mercury expands uniformly with heating, it is also often used in thermostats, which are heat-operated electrical switches.

Mercury is also used to make amalgams. An amalgam is a metal alloy produced by mixing mercury with another substance. Mercury's ability to form amalgams is important in the extraction of gold and silver from their ores. More familiarly, an amalgam containing silver may be used for dental fillings. However, because of concerns about the toxicity of the metal, dental fillings are now often made from other materials.

▲ *This is a photograph of the mineral cinnabar. Cinnabar is an ore of mercury and has a distinctive red color. It is also used to produce the red pigment vermillion. Mercury is most often produced from cinnabar, by roasting it at high temperature.*

There are many other uses of amalgams and other mercury compounds. These range from controlling the spread of fungi in crops to acting as detonators in explosives. Mercury and most of its compounds are, however, highly poisonous and some of their earlier uses, such as in antiseptic creams, have been discontinued.

See also: ALLOY • BAROMETER • DENTISTRY • EXPLOSIVE • THERMOMETER

Mercury, planet

Mercury is the planet nearest the Sun and the smallest planet in the solar system except for Pluto. It is a rocky planet, with a surface that resembles the surface of the Moon. Mercury can sometimes be seen from Earth as a faint evening or morning star.

Mercury travels around the Sun at an average distance of about 36 million miles (58 million kilometers). It is too close to the Sun for astronomers to see it clearly. Even the best telescopes show it merely as a pinkish disk with faint dark markings.

Mercury has been known since at least the time of the Sumerians about 5,000 years ago. Ancient Greek astronomers gave the planet two names—Apollo for when it appears as a morning star and Hermes for when it appears as an evening star. However, they knew that Apollo and Hermes were the same. Greek astronomer Heracleides (c. 390–c. 322 BCE) even believed, correctly, that the planet orbits the Sun and not Earth.

Hermes was the swift-winged messenger of the Greek gods, so that seemed a fitting name for this planet, which moves through the sky faster than any other. The Romans called the planet Mercury, and this has remained its name.

Mariner discoveries

Not until the *Mariner 10* space probe visited Mercury between 1974 and 1975 did astronomers get a clear look at the planet's surface. *Mariner 10* revealed that billions of years of constant battering by meteorites had left the surface more deeply dented with craters than the Moon, some more than 100 miles (160 kilometers) across. All that can be seen on the planet's surface are cliffs hundreds of miles long, yellow dust, and vast, empty basins. In fact, Mercury's surface resembles the lunar landscape. The main difference is that Mercury has far fewer large flat mare ("sea") areas similar to those on the Moon, and that the maria (*singular,* mare) on Mercury are much lighter in color.

◀ *This is one of the first ever close-up images of Mercury, taken by the **Mariner 10** space probe as it sped past the planet on March 29, 1974. The picture was taken from about 37,300 miles (60,000 kilometers) away.*

▶ *The curved shadow on the left of this picture is one of the dramatic cliffs, or scarps, on Mercury revealed by the **Mariner 10** space probe. This one is about 220 miles (350 kilometers) long.*

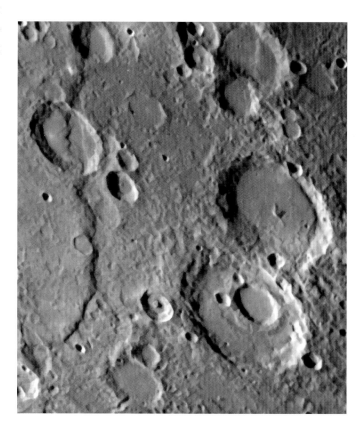

The craters on Mercury's surface were given all kinds of cultured names after the *Mariner 10* voyage, including those of composers Bach and Beethoven and writers such as Homer and Tolstoy.

The most prominent feature of Mercury is a huge basin, Caloris Basin, which covers nearly one-tenth of the planet's surface. Its diameter is 800 miles (1,300 kilometers), which is more than one-quarter the diameter of the entire planet. The crater floor is about 5½ miles (9 kilometers) below the average height of the surface of Mercury, similar to the deepest ocean floors on Earth.

A ring of mountains, more than 1 mile (1.6 kilometers) high, encircle the basin. Parts of the basin seem to have been flooded with molten rock and may contain old volcanic craters. The basin was probably formed by the impact of a giant meteorite. The impact undoubtedly led to huge "mercury-quakes," which caused the jumbled, cracked landscape called weird terrain.

Mercurial volcanoes

At first glance, the *Mariner 10* images suggested that Mercury was basically an unchanging world, shaped only by impact craters. But a second look at the *Mariner 10* data by scientists from the National Aeronautics and Space Administration (NASA) has shown that there has been much more volcanic activity than was previously imagined. All over the surface are smooth plains, at least some of which are probably the result of lava flows from ancient volcanoes. In other places, such as the 200-mile (320-kilometer) wide Homer crater, there are signs of the debris from explosive volcanic eruptions, called pyroclastic deposits.

Radar data also hint that Mercury, like the Moon, has deposits of frozen water in permanently shaded areas at the poles. Unlike the Moon, which has ice only at its south pole, Mercury has ice at both poles. The ice caps are not entirely water ice, though. Much of the ice is sulfuric acid (H_2SO_4).

Mercury's atmosphere

Mercury has a diameter of about 3,010 miles (4,840 kilometers), about one-third the diameter of Earth. Being so small, it is unable to retain much of an atmosphere, just a few wisps of sodium vapor. The slight "atmosphere" detected by *Mariner 10* probably consists of particles trapped in its magnetic field, which may indicate that the planet has an iron core like Earth's. This would also account for Mercury's high relative density (5.4). It is very similar to the relative density of Earth.

Why is Mercury dense?

One possible explanation for the high relative density of Mercury is that a major collision with another large body, such as a meteoroid, knocked away part of Mercury's outer crust, leaving an oversized core. The large eccentricity of the planet's orbit (the degree to which it departs from a circular shape) and the unusually large inclination of its orbit (the degree to which it is tipped away from the common plane of the other planetary orbits) might have been caused by such a severe collision.

A second possible explanation is that because of Mercury's proximity to the Sun, the more volatile elements never condensed from the solar nebula—the cloud of gases that surrounded the newly formed Sun. A third explanation lies in the earliest stages of Mercury's development, when it was growing by attracting and fusing with objects that collided or came close to it. It is possible that lightweight objects were less likely than heavy objects to fuse with the young planet, owing to the high gas densities and rapid movement of objects close to the Sun.

▲ **This is an artist's impression of the Messenger probe, launched in July 2004. Messenger is the first probe to visit Mercury since Mariner 10. It is hoped it will make significant new discoveries about Mercury.**

Mercury's long days

The temperature on Mercury soars to more than 750°F (400°C) during its "midday," partly because the planet is so close to the Sun. The Sun looks three times as big in Mercury's sky as it does from Earth. The scorching temperature is also caused by Mercury's long exposure to sunlight. Earth rotates on its axis once every 24 hours, and so the Sun rises and sets on Earth once every 24 hours, as parts of Earth's surface turn around to face the Sun and then swing away again. Mercury, however, spins very slowly, turning only once every 59 Earth days.

Since Mercury speeds around the Sun in just 88 days, the planet is spun around to face the Sun from the other side almost as fast as the sunlit side swings away again. So once the Sun rises on Mercury, it takes a very long time to set again. In fact, it sets only after 176 Earth days, or two Mercury years, because in that time the planet has orbited the Sun twice.

The combination of Mercury's slow rotation and rapid orbit can produce some strange effects. Twice during its orbit, Mercury gets very close to the Sun

▶ *This is a false-color photograph of Mercury's surface taken by the Mariner 10 probe in 1974. The picture was compiled from many separate photographs and reveals the planet's moonlike surface.*

and begins to travel much faster. The result is that for a few hours, the Sun actually appears to go backward in Mercury's sky. At some points on Mercury's surface, the Sun appears to rise, then grow larger, then travel backward for a few hours, then travel forward, getting smaller until it finally sinks. Meanwhile, the stars would be moving three times faster across the sky.

The search for Vulcan

Mercury's orbit around the Sun is far from circular. It is a very long ellipse. At its nearest, called its perihelion, Mercury is 28 million miles (46 million kilometers) from the Sun. At its farthest, called its aphelion, it is more than 43 million miles (70 million kilometers) away. The perihelion point is not fixed but gradually moves around the Sun, year by year. Astronomers could explain most of this perihelion shift as the effect of the gravity of the other planets in the solar system. However, this still left a little bit of the shift unexplained. So in the nineteenth century, astronomers began to think there might be a tiny undiscovered planet nearby, which they called Vulcan, that was causing the shift. Astronomers now know there is no mystery planet named Vulcan. Instead, the tiny shift in Mercury's perihelion turned out to be the first real proof of the theory of special relativity, put forward by German-born U.S. physicist Albert Einstein (1879–1955), who suggested that space itself is slightly curved by the gravity of a massive object like the Sun. The curvature of space bends light slightly and so makes Mercury appear to be in a slightly different position from where it actually is.

Viewing Mercury

Mercury can sometimes be seen with the naked eye. Although it is impossible to see any detail, even with binoculars, it is easy to tell it is a planet because its light is so much steadier than the stars. It only twinkles when it is close to the horizon.

Sightings are rare, however. Several famous scientists, including Polish astronomer Nicolaus Copernicus (1473–1543), lived their entire lives without catching a glimpse of the planet. It may be quite bright, but it is so close to the Sun that it can only be seen clearly on a few days each year, usually in the spring and the fall. Even then, it can never be seen high in the sky or during the night. Like Venus, Mercury is an evening or morning star, and it appears as a pale glimmer just above the horizon at dusk or dawn.

See *also:* EARTH • MARS • SOLAR SYSTEM • SPACE PROBE • SUN

Metabolism

All the chemical processes taking place inside a living cell or organism are called metabolism. Some processes involve the breakdown of chemicals to provide energy. Other processes use energy to build up complicated body chemicals. Metabolism occurs in the cells of the body and involves two main processes—catabolism and anabolism.

All life depends on a complex array of chemical processes that both create new materials and dispose of unwanted materials. Energy is needed to make both these processes happen. *Metabolism* is the word used to describe all these processes together. It comes from the ancient Greek, meaning "throw differently," and can be roughly translated as "changing." Metabolic processes occur all the time, building chemicals up and breaking them down. There are two kinds of process at work—catabolism and anabolism.

Catabolism

In essence, catabolism is the breakdown of complex chemicals into simpler ones to release energy. Some of this energy is used to make muscles move and keep the body warm. Most is involved in making anabolic processes working.

Catabolism works by breaking down energy-rich chemicals the body gets in nutrients such as amino acids, fatty acids, and starch and sugar. Starch and sugar are carbohydrates. They get their name because they are made mainly of carbon and hydrogen. Carbohydrates are the body's main source of energy. When carbohydrates are eaten, the body converts them into a form of sugar called glucose, which is then distributed to the body cells by blood in the circulatory system.

Cells catabolize glucose to release energy. In simple terms, this process, called cellular respiration, means splitting the carbon and

▼ *The cells in this athlete's muscles are catabolizing rapidly, converting sugar into energy at a huge rate to keep her moving. Burning all this sugar requires lots of oxygen, so she needs to breathe deeply.*

hydrogen in glucose to release energy. The hydrogen joins with oxygen to make water and the carbon joins with oxygen to make carbon dioxide.

Glucose catabolism involves two steps. It begins with glycolysis. Glucose is first broken down into pyruvic acid and releases a little of its energy. The second step, called the Krebs cycle, only occurs if oxygen is present. First, the pyruvic acid is converted to acetyl-coenzyme A (acetyl-CoA). Then the acetyl-CoA combines with oxygen in a series of reactions to yield carbon dioxide (CO_2) and water (H_2O). Carbon dioxide is poisonous and is removed from the body when a person exhales.

At the same time, large amounts of energy are released and stored in a chemical called adenosine triphosphate (ATP). This energy is available for use in other body processes. If there is not enough carbohydrates to provide needed energy, however, the body starts to break down fats and sometimes even proteins.

Fatty acids can also provide energy, but they are catabolized in a different way, changing first to acetyl-CoA, then going through the Krebs cycle. Amino acids can also provide energy by going through the Krebs cycle, but first they must be converted into the right form in the liver.

▲ *When sleeping, the body's metabolic rate drops to a minimum, called the basal metabolic rate (BMR). A woman's BMR is slightly lower than a man's, because she has more body fat to keep heat in.*

Anabolism

Anabolism is the reverse of catabolism. It is the process the body uses to build up complex body chemicals from simpler ones. This is the way the body makes structural proteins, which are used to repair tissues and grow new ones, and functional proteins that perform particular tasks in the body, such as the enzymes that speed up chemical reactions and the hormones that trigger many body processes. Anabolic processes also make lipids— fatty substances the body uses to bind cells together.

Metabolic rate

The speed at which chemicals are built up or broken down is controlled by hormones, particularly those produced by the thyroid gland, such as thyroxine. The rate of metabolism is also influenced by factors such as diet and temperature.

A person's metabolic rate is the amount of energy he or she uses up in one hour. This is usually measured in large calories, and one large calorie equals 1,000 small, or gram, calories (1 kilocalorie).

When running, a sprinter's metabolic rate increases dramatically, and a large amount of energy is used up in a very short time. Energy comes to the muscles in the form of ATP, which is made when glucose is broken down during respiration. The amount of energy supplied to the muscles depends on a number of factors, particularly the amount of oxygen available.

Metabolic rate includes two ratings. A person's basal metabolic rate (BMR) is the amount of energy used up per hour when he or she is completely at rest. It is the energy needed to keep the body working without any muscular activity. The BMR of a man is usually slightly higher than that of a woman because women tend to weigh less than men and have more body fat to keep heat in. An average-sized man has a BMR of about 65 kilocalories per hour, and an average-sized woman has a BMR of about 55 kilocalories per hour.

When a person is using his or her muscles, a second part is added to the BMR. The greater the muscular effort, the more energy is used up. For example, a man may use up to 90 kilocalories an

DID YOU KNOW?

Metabolism produces many waste chemicals that have to be excreted from the body. This excretion usually occurs by sweating or urinating.

hour just sitting down, 320 kilocalories an hour when walking, and 600 kilocalories an hour when running. A woman might use 70, 180, and 420 kilocalories an hour, respectively, performing the same activities.

How food is used in the body

Most energy comes from carbohydrates but, if necessary, fats and proteins can also be used to supply energy. Most proteins, together with some carbohydrates and fats, are used for building and repairing cells. As enzymes, proteins enable chemical reactions that cause growth in cells to take place rapidly, particularly under neutral conditions; that is, when the organic system is neither particularly acidic nor basic.

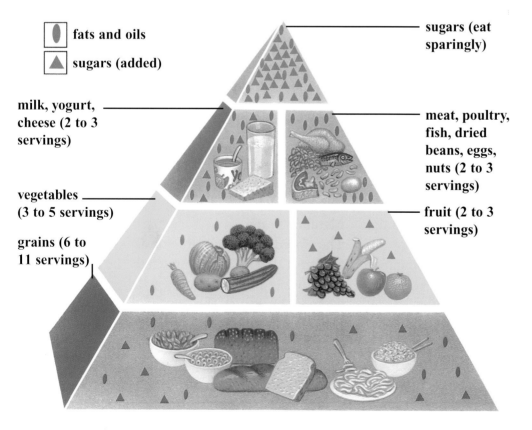

fats and oils

sugars (added)

milk, yogurt, cheese (2 to 3 servings)

vegetables (3 to 5 servings)

grains (6 to 11 servings)

sugars (eat sparingly)

meat, poultry, fish, dried beans, eggs, nuts (2 to 3 servings)

fruit (2 to 3 servings)

◀ *This pyramid shows how many daily servings a person should choose from each food group. For a balanced diet, a person should eat food from each group. In 2005, the U.S. Department of Agriculture released the new MyPyramid symbol developed to convey the government's guidelines for a healthier diet and lifestyle. See MyPyramid.gov for further details.*

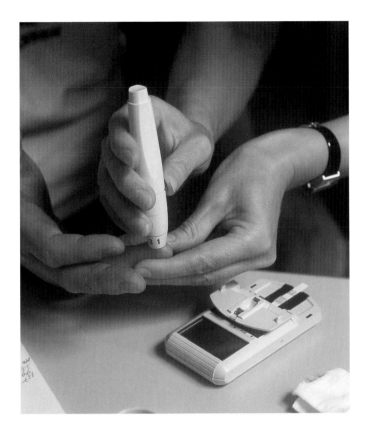

◄ *A mother tests the blood sugar levels of her diabetic child. Diabetes is a metabolic disorder in which the pancreas does not make enough of a hormone called insulin. The insulin deficiency impairs the body's ability to metabolize glucose.*

Efficient metabolism, which is essential to normal body functioning, requires a balanced and regular supply of nutrients. A bad diet, overeating or undereating, the heavy consumption of alcohol, and even certain drugs can interfere with the metabolic system and cause potentially lethal problems. The lack of just one enzyme can block a chemical pathway and cause too much of a particular chemical to build up in the body.

Hormones control processes such as growth and the burning of energy reserves. Insufficient production of hormones can cause metabolic problems. Diabetes mellitus, for example, is a metabolic disorder that results when the pancreas does not produce enough insulin—the hormone that enables body cells to metabolize glucose. The glucose then accumulates in the blood. To get energy, fat stores are broken down, but these cannot be catabolized, and unprocessed products poison the body. People who have diabetes must be careful to eat the right foods. Exercise is also important. About half of all people with diabetes also require insulin injections.

Most metabolic problems result from an imbalance between the food taken in and energy used up. An active person, or someone with a high metabolic rate, needs a high food intake. This is also true of most growing children. If insufficient food is taken in, then the person will lose weight. On the other hand, an inactive person, or one with a low metabolic rate, needs much less food. If excess food is eaten, it is not used up. Instead it turns into fat. The accumulation of fat in the body may lead to serious health problems, such as obesity and associated risks such as heart disease.

Proteins and fats

Proteins are assembled in various combinations from 20 different amino acids. Just as the 26 letters of the alphabet can be assembled to form words of various lengths and meanings, a few, or many, of the 20 different amino acids join up to form different proteins.

Fats are stored in the form of adipose (fat) tissue, which provides the body with insulation and acts as a depot for energy use. Excess carbohydrates and proteins can also be stored as fat. A person who does not eat enough food uses up all stored reserves and loses weight, while a person who eats more food than necessary stores the excess chemicals as fat. Children derive their energy from consuming food. They need energy not only for muscle activity but also for growth.

Metabolic disorders

By assisting and speeding up chemical reactions, enzymes influence chemical conversions so that required substances are made available to body cells. Some metabolic disorders are caused by deficiencies of certain enzymes.

See also: CELL • ENDOCRINE SYSTEM • ENZYME • EXCRETORY SYSTEM • EXOCRINE SYSTEM • MUSCULAR SYSTEM • NUTRITION • PROTEIN

Metal

Metals are shiny, mostly solid materials, such as gold, iron, and tin, and are some of the most common elements on Earth. They are also useful for making things because, although they are strong, they can be bent, hammered, or molded into shape. People rely on metals to make everything from teaspoons to jet engines.

Together, metals form one of the most important of all groups of substances. In fact, all elements can be divided into metals and nonmetals. Of the 92 elements that occur naturally on Earth, more than 60 are metals, such as copper, gold, and tin. A further 10 have properties that place them halfway between metals and nonmetals. Just 20 are nonmetals, such as chlorine and sulfur.

Some metals are very common, such as calcium, iron, magnesium, potassium, and sodium. The most common metal of all in the ground is aluminum. Deep inside Earth, the core is made almost entirely from iron, but this is inaccessible. Many metals, though, such as gold and platinum, are much rarer and so are often very valuable.

The properties of metals

Each metal has its own properties, but most pure metals are solids at standard room temperature and pressure. The exception is mercury, which is liquid down to about –38°F (–39°C). Similar to all the metals, however, mercury is shiny. Some metals can be polished so that they reflect light, but the surface may have to be scraped clean to reveal this shininess. Many metals, such as aluminum, become dull and tarnished if exposed to the air, because they react with chemicals in the air. Some metals even change color. The green tarnishing of copper is called verdigris. Iron will become brown and rusty if exposed to air and water for long enough.

Most metals are rigid and strong. They have what is called "tensile strength," which means they can support a heavy load. For this reason, large structures, such as building frames and bridges, are often made of metal. However, metals can sometimes also be bent without breaking. They are said to be malleable, which means they can be hammered or rolled into different shapes. They are also said to be ductile, which means they can be pulled out to form wires.

Metals also conduct electricity and heat well. Among the best conductors of all are copper and silver. Electric wires are usually made of these metals. Copper also conducts heat well, so the best saucepans are often made of copper. Many metals even conduct sound well, and they ring when struck. Bells are therefore made of metal. Lead, though, is so soft and dense that it absorbs sound and will not ring.

Metals vary widely in their reactivity to other chemicals. Lithium, potassium, and radium are all highly reactive. Gold and platinum are very unreactive. Gold can lie in the ground for thousands of years and still be dug out looking almost as shiny as if it were new.

The structure of metals

Although it is not immediately obvious, metals are actually crystalline—that is, they consist of crystals. A block of metal consists of many crystals, which are usually either cubic or hexagonal in shape. In turn, the crystals consist of many tightly packed atoms. It is this close packing of the atoms that makes metals dense and heavy.

Many chemicals form molecules consisting of just a few atoms. Metals atoms, however, rarely form molecules in this way. Instead, they form what are called giant structures. These are huge lattices or frameworks of neatly arranged atoms held together by what are called metallic bonds.

▲ *Most large modern buildings, from apartment houses and skyscrapers to factories, could not be built without their framework of immensely strong steel girders.*

▶ *The ductile strength of many metals, such as steel, means they can be drawn out to make long, thin cables such as these. These metal cables are much stronger than any natural fiber rope.*

It is because these metallic bonds are so rigid and the structure so firm that metals are normally so strong. The atoms are held together because they are surrounded by a mass of electrons. At any instant, two neighboring atoms may have opposite electrical charges and will attract each other.

Although atoms are generally arranged in a regular pattern, there are always defects, known as dislocations, where groups of atoms can slide over each other. It is these dislocations that allow metals to bend and to be hammered or stretched into different shapes. Too many of these defects, however, make the metal weak.

Usually, electrons are attached to a particular atom or are shared directly between two atoms in a molecule. In a metal, some excess electrons become entirely detached from their atoms and are able to move freely. It is these free electrons that make metals such good conductors of heat and electricity. Metals are good conductors of heat because heat energy is easily passed from one atom to the next. Electricity flows well through metal because of the movement of the free electrons. These electrons also absorb and re-emit light easily, which is the reason that metal is opaque and shiny.

DID YOU KNOW?

Metal detectors are in use all the time. When traveling by airplane, passengers pass through an arch-shaped metal detector before they board a flight. Many shops tag their goods with metallic strips and have modified metal detectors by entrances and exits to alert security guards to theft. More recently, entrances to schools and other public spaces have metal detecting archways or barriers to detect the presence of dangerous or illegal weapons.

Metal groups

Most metals melt at high temperatures and are hard and dense. One small group of metals, called the alkali metals, is very different. The alkali metals comprise lithium, sodium, potassium, rubidium, and cesium. All the alkali metals are soft, have low melting points, and are low in density. Lithium, sodium, and potassium actually float on water. Lithium is the lightest solid element of all. All the alkali metals are also so reactive that they are

ALLOYS COMPARED TO PURE METALS—STRENGTH FROM DIVERSITY

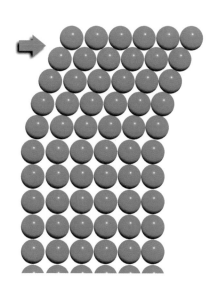

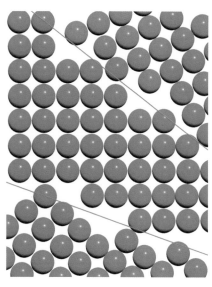

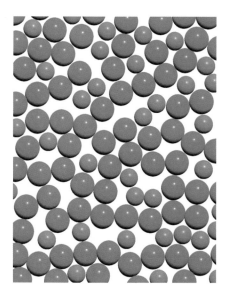

▲ *Pure metals are relatively soft, since their crystal structures are easily deformed.*

▲ *Dislocations (indicated by the lines) allow metals to be shaped, but too many make it weak.*

▲ *The different sizes of atoms in alloys make metals stronger but still easy to shape.*

dangerous, reacting even with water. For example, even a small amount of potassium fizzes and glows violently if just dropped in water.

Another metal group is the alkaline earth metals. They include beryllium, magnesium, calcium, strontium, and barium. Similar to the alkali metals, they are fairly reactive, though not as much. Barium, though, reacts so violently with both water and oxygen that it has to be stored under oil.

A third group of metals, and the largest by far, is what is called the transition metals. Transition metals include both very common metals, such as copper, iron, and zinc, and very rare metals, such as iridium, scandium, and vanadium. They have all the typical qualities associated with metals, but they are especially tough, which makes even the rarest of them useful. Iridium is used in heart pacemakers, while niobium alloys are used in rockets. They also have the highest melting points of all metals—tungsten having the highest melting point of any metal at 6170°F (3410°C).

Metalloids

Transition metals get their name because of their position in the periodic table of chemical elements. The alkali and alkaline earth metals are found on the left of the table, and nonmetals on the right. The transition metals fall in between.

So, also, do Groups III, IV, and V. The heavier elements in these three groups, such as aluminum, are metals, but the lighter ones, such as carbon, are nonmetals. Many elements in these three groups have properties that fall between metals and nonmetals and are called metalloids. Arsenic is a shiny metal, but some of its chemistry is similar to that of the nonmetals. The metalloids include silicon, germanium, arsenic, antimony, and selenium. Some are semiconductors—substances that conduct electricity only at high temperatures and that act as insulators at low temperatures.

Alloys

Many pure metals can be surprisingly soft and weak. It is only when they have traces of other substances added that they become strong. So most

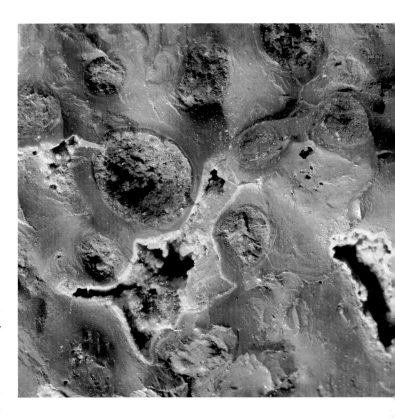

▲ *This photograph is a close-up of bauxite, which is the main ore of aluminum. This sample is from Beaux in France, where aluminum was first identified by the French mineralogist P. Berthier in 1821.*

metals are mixtures in which pure metals are combined with traces of other substances to make them stronger or give them certain qualities. These combination metals are called alloys.

Alloys have been used for thousands of years. Copper was the first pure metal to be used, but it was soft, and so copper knives were never sharp. Then, about 6,000 years ago, people in the Middle East discovered that by adding tin to molten copper, they could make it much stronger and less prone to turning green. This alloy was called bronze and was so widely used for everything from bowls to axe-heads that this period is called the Bronze Age. Another old and widely used alloy is brass, which is an alloy of copper and zinc used by the Romans. Brass has a color almost like gold.

Steel and modern alloys

The most important alloy of all, however, is steel. Steel is an alloy of iron that has been known for more than 2,000 years. Basic steel is made not by

▲ *By heating it until it melts to a liquid, metals, such as this molten iron in an ironworks, can be poured into a mold of almost any shape. Iron melts at a temperature of 2795°F (1535°C).*

adding substances to iron but by taking them out. Iron typically contains a little (about 3 percent) carbon, which makes it hard but rather brittle. Steel is made by taking some of this carbon out to make it tougher and less brittle. There is actually a huge variety of alloy steels, each made by adding different substances to give particular qualities. For example, stainless steel that does not rust is made by adding traces of chromium.

In recent years, many new alloys have been made for automobiles, aircraft, electronic equipment, and spacecraft. For example, many aircraft bodies include a very light, strong alloy of aluminum and copper called duralumin.

Metal ores

Metals may be very common in Earth's crust, but they only rarely occur in pure form. They nearly always occur in combination with other chemicals in minerals called ores. There are about 100 of these altogether, each containing different metals or combinations of metals. Iron is found in ore minerals such as hematite, limonite, magnetite, and siderite. Copper is found in ores such as chalcopyrite. Lead is found in the ore galena.

To get the metal, the ores must not only to be dug from the ground, but they must go through a refinement process to extract the pure metal. While with some ores, such as copper, this just involves heating, with others the refinement process is long and complex. Although aluminum is the most plentiful metal in Earth's crust, it was not even known, let alone exploited, until 150 years ago.

The magma that wells up from Earth's interior is rich in metals. Metals that combine readily with sulfur—such as copper, iron, and zinc—form in hot, sulfurous magma under volcanoes, and they form ore crystals as the magma cools and solidifies. The famous aluminum, copper, and nickel ores in the Sudbury area of Canada were formed in this way. Metals that combine readily with silica, such as beryllium, potassium, and uranium, form in cracks where super-hot, silica-rich liquids bubble away from the edge of cooling domes of magma underground. Where the hot fluid is forced at high pressure into tiny cracks called veins, it forms copper, gold, and molybdenum.

Other ores form in sediments created as rocks are broken down by the weather, leaving a metal-rich residue. In the tropics, the aluminum ore bauxite forms in huge deposits in this way. Metal-rich residues may be washed into the sea and collect on the seafloor. That is how the rich lead and zinc layers formed at Australia's Broken Hill, and copper layers in Zambia's Copper Belt.

▲ *The possibility of discovering buried treasure has turned metal detecting into a popular pastime in Europe, where ancient metal artifacts have often remained hidden for thousands of years.*

Sometimes, rivers wash weathered rock into shoals of sand that contain small quantities of gold, platinum, and other precious metals. The metals are extracted from "placer deposits" by washing away the sand and gravel.

Discovering metals

The first metals discovered were the least reactive of metals and the easiest to isolate from their ores, such as copper, gold, and lead, which are often found almost pure. Those that had to be extracted using heat, such as iron and zinc, took longer to find, but even these were in widespread use 3,000 years ago. Reactive metals took much longer to discover. Potassium was found by English scientist Humphry Davy (1778–1829) in 1807. Davy also found calcium, magnesium, and sodium. Aluminum is so difficult to extract from its ore, called bauxite, it was not discovered until 1821, even though it is the most common of all metals.

Metal detectors

The discovery and extraction of metal deposits was helped by the invention of metal-detecting technology. Late in the nineteenth century, Scottish inventor Alexander Graham Bell (1847–1922) designed the first crude metal detector. It was not until 1931, however, that portable metal detectors became a commercial success. A metal detector is used to find hidden metal objects. Archaeologists, the military, police, prospectors, and treasure hunters all use these lightweight instruments.

Metal detectors work on the principle that a metal object causes a change in a magnetic field. If an alternating magnetic field (one that constantly changes strength and direction) is set up near a metal object, small electrical currents are created in the object. The metal detector picks up the slight changes in the magnetic field created by these electrical currents.

See also: ATOM AND MOLECULE • ELECTROLYSIS • ELEMENT, CHEMICAL • GEIGER-MÜLLER DETECTOR • MASS AND WEIGHT • MINERAL

Metallurgy

Metallurgy is a branch of chemistry that looks at ways in which metals are extracted and their properties changed to fit particular uses. It requires knowledge of the chemical and physical behavior of metals as well as the engineering methods used in industry.

Metallurgy is one of the oldest sciences known to humankind and has guided the progress of technology over the last five millennia. Ancient peoples learned early metallurgy by trial and error. They also believed it to be part of the art of alchemy, which sought to convert base metals, such as lead, into precious metals, such as gold. Modern metallurgy is an exact science that looks at the atomic, chemical, and physical properties of metals.

Metallurgy covers many subjects, so it is often divided into different areas. Extractive metallurgists look at the ways in which metal compounds are separated from their ores—rocks that contain metal compounds as they are found in their natural form on Earth. They are also involved in refining, which is the process used to free metals from impurities, and in the manufacture of alloys (mixtures of metals). Chemical and physical metallurgists examine the structure and properties of metals to find ways of converting them into useful products. Metallurgists also investigate mechanical processes, such as how to cast, forge, or

▼ *This smelter at the Cananea copper mine in Mexico produces about 55,000 tons (50,000 tonnes) of refined copper each year. The copper ore is smelted in high-temperature furnaces to produce blister copper, which is 99 percent pure. The blister copper is then further refined in a process called electrolysis.*

▶ *This illustration of medieval metalworkers in Germany was published in Agricola's monumental work* **De Re Metallica** *(1556).*

roll metals. A related subject is powder metallurgy, which is used to manufacture items by pressing metal powder at extremely high temperatures.

The history of metallurgy

The first metals known to ancient civilizations were almost certainly copper, gold, and silver, all of which are found naturally in riverbeds. In the fourth millennium BCE, artisans in central Europe were casting copper into cutting tools and weapons. By 2500 BCE, the ancient Egyptians were using kilns to extract copper and gold from their ores. They had also learned how to cast and forge metal objects. These methods were later used on bronze (an alloy of copper and tin) and then iron, giving rise to the Bronze Age and Iron Age.

In Greek and Roman times, between about 500 BCE and 500 CE, metallurgists had learned how to make alloys such as brass (a mixture of copper and zinc) and steel (a mixture of iron and carbon). They had also discovered lead and mercury. However, little progress followed as Europe entered the Middle Ages. Metallurgy enjoyed renewed popularity in Europe in the sixteenth century thanks to the work of German scholar Georg Bauer (1494–1555), better known by his Latin name, Agricola. Agricola wrote the influential book *De Re Metallica* (1556) in which he described techniques such as ore extraction, refining, and smelting.

DID YOU KNOW?

Lithium is the lightest (least dense) metal, while iridium and osmium are the heaviest (most dense) metals. Chromium is the hardest metal. Mercury is a liquid at room temperature and is therefore the softest metal. Silver is the best conductor of heat and electricity. The most common metal on Earth is aluminum, while iron is the most widely used in industry.

Extracting metals from ores

To extract a metal from its ore, the useful material must first be separated from the rock in which it is found. Some form of chemical reaction is then used to obtain the pure metal. The process of extracting a metal from its ore is called refining.

There are several methods for concentrating metal ores. Panning crushed ore in a stream is one of the oldest and simplest methods. The moving water carries away the lighter rocks, leaving behind the heavier ore. Some modern methods still use running water to separate ore, such as in tin refining. However, techniques such as sieving or using magnets are more popular. Frothing air through a mineral sludge is also widely used, since some ores stick to the air bubbles. The mineral-rich bubbles can then be skimmed off the surface of the sludge. This process is called flotation separation.

The type of reaction used to extract a metal from its ore depends on the chemical composition of the ore. Iron ores such as hematite (largely Fe_2O_3) and magnetite (largely Fe_3O_4) are usually smelted. In this process, the ore is mixed with carbon and heated to very high temperature. The carbon then pulls oxygen atoms off the iron atoms, producing carbon dioxide gas (CO_2) and molten iron. Copper and lead ores often contain sulfur instead of, or as well as, oxygen. In this case, the ore is roasted in air to replace the sulfur with oxygen before smelting.

Physical properties of metals

All metals are shiny and are good conductors of heat and electricity, but they can otherwise take a wide range of physical properties. Some are soft, such as gold and lead, and can easily be hammered into shape. Others are very hard, such as iron and titanium, and do not easily distort. Other important properties include brittleness, density (mass per unit volume), and the melting and boiling points.

Physical metallurgy looks at these features of metals and how they may be changed to fit particular uses. For example, annealing is a form of

▲ A technique called flotation separation is used to concentrate tungsten ore. First, the ore is crushed and mixed with water and chemicals in a flotation tank. When air is blown through the mixture, mineral-rich bubbles collect at the surface. The surface can then be skimmed to increase the concentration of the ore.

> **DID YOU KNOW?**
>
> Tungsten metal has the highest melting point of all the metals (6152°F or 3410°C), so it is extremely difficult to melt and pour into molds. As a result, chemists shape tungsten into useful objects, such as the filaments for electric lightbulbs, using the process of powder metallurgy.

heat treatment that softens a metal. The metal is heated to a high temperature and then allowed to cool slowly. Metalworkers can also harden a metal by heating it and then quickly quenching (cooling) it in water. This results in a hard but very brittle metal. Hammering hot metal makes it stronger.

Another way of making metals with particular properties is to mix an alloy. Alloys consist of a mixture of metals or a metal and a nonmetal. For example, the main ingredient in stainless steel is a mixture of iron and chromium, called ferro-chromium. The physical properties of the alloy depend on the amounts of their individual ingredients and how they are made.

Chemistry and metallurgy

Different metals can also have a wide range of chemical properties. Some are very reactive, such as calcium or sodium, both of which fizz violently in contact with water. Others are inert (unreactive), such as the noble metals gold and silver, and can be found naturally in their pure form. It is also important for metallurgists to know the type of compounds that metals form with other elements.

A large part of chemical metallurgy looks at how to modify metals to protect them against corrosion (attack by rust or chemicals). Steel corrodes when it comes into contact with water and oxygen atoms in the atmosphere. However, the addition of chromium to stainless steel stops the rusting process. The chromium atoms near the surface of the stainless steel combine with oxygen atoms in the atmosphere, forming a thin, protective chrome-containing oxide, called the passive film. The sizes

◀ *The glowing orange rod in this picture is a sintering furnace, which is used to form metal objects from powdered metal. The metal powder is heated in a mold for several hours at a temperature that is below the melting point of the metal. This causes the powder to fuse and form a solid object.*

of chromium atoms and the oxide molecules are similar, so they pack tightly on the surface of the alloy, forming a stable layer only a few atoms thick.

Chemical metallurgists also try to find the most efficient ways of extracting metals from their compounds. Many reactions are complex; they may only proceed in the presence of particular gases or under certain pressures. By figuring out the optimum conditions needed for the reaction to proceed, a chemical metallurgist can save a refinery a lot of money.

Powder metallurgy

Metal objects are usually made in one of two ways. The metal may be melted and then poured into a mold (called casting). Alternatively, the metal may be compressed or stamped into the required shape. However, many metals cannot be made easily using either of these methods. Some have high melting points; others shrink as they cool. A metal may also be weakened using either of these methods.

In powder metallurgy, also called sintering, the metal is first ground into a fine powder. If the object is to be made from an alloy, then the powders of the various metals are mixed together. The powder is then compressed in a machine, forming a compacted shape of the desired object. Finally, it is placed in a high-temperature sintering furnace so that the powder fuses to become a solid. The compression and heating may be completed in one operation called hot pressing.

Many items are made using powder metallurgy. The tips of drills consist of a mixture of very hard metals and can only be made using powder metallurgy. The method can also be used to make large numbers of small metal articles cheaply and accurately, such as the parts for sewing machines.

See also: ALLOY • CASTING • CHEMICAL REACTION • CORROSION • METAL • METALWORKING • MINING AND QUARRYING

Metalworking

The skill of metalworking dates back thousands of years. Early craftspeople hammered metal objects into shape in a forge. Modern techniques include rolling and electropolishing, and huge machines are now used to shape metal items weighing hundreds of tons. With the advent of computer technology in the twentieth century, metalworking became highly automated.

▲ *When a blacksmith heats a bar of metal and shapes it into a horseshoe by beating it with a hammer, he or she is forging. In modern industrial plants, large furnaces have replaced the blacksmith's forge, and powerful machines have replaced the hammer.*

The art and science of metalworking is a skill that has helped humanity reach the modern technological age. It allowed ancient peoples to make weapons for war and tools for agriculture. Moreover, many of the necessary items of a civilized society are crafted from metal. Without metal coins, for example, it would be impossible to have modern economies and taxes.

Different metals have different properties that make them suitable for different applications. Some are very hard and strong, such as iron, making them perfect for weapons or tools that need to stay sharp. Others are soft and malleable, such as gold and lead, which means that they can be easily shaped into items such as jewelry and pipes. All metals can be polished to a beautiful reflective luster. Noble (unreactive) metals, such as gold and silver, keep their luster longer because they do not easily react with oxygen in the atmosphere.

Early metalworking

Humankind has worked metal into useful and decorative items since the dawn of civilization some eight or nine millennia ago. Archaeologists have found copper ornaments in Mesopotamia (present-day Iraq) dating back to about 8700 BCE. Metal for such early objects came from native (natural) sources, such as copper or gold in riverbeds and iron from meteors.

With the discovery that pure metal could be smelted from mineral-rich rocks, metal became more plentiful about 4000 BCE. This discovery prompted two periods in Earth's history—first the Bronze Age and then the Iron Age. The earliest technique for shaping metal was to hammer heated metal into shape. The object could then be decorated with engravings or inlaid with precious stones or other metals. Skilled artisans used these methods to make beautiful artifacts, such as elaborate chalices and delicate jewelry.

The mechanization of metalwork began with the Industrial Revolution in eighteenth-century Britain. Factories manufactured huge amounts of iron and steel to make bridges, railings, steam engines, and later ships and railroads. New techniques for shaping metal, such as pressing and rolling, became common. This technology created many features of today's modern world, such as automobiles, skyscrapers, and mechanized warfare.

Forging metal

Inside all solid metals is a pattern of grains. Beating the metal changes the grain pattern to make the metal tougher. The grains become finer and can slip over each other, so the metal is less likely to break. This process is called forging. Forged items, such as the machine parts for engines, propeller shafts for ships, and turbine blades, can withstand heavy wear. Most modern forges use huge forging hammers in a method called drop forging. The heated metal to be shaped, called the billet, is placed between two shaped tools called dies. When a hammer falls onto the dies, it presses them together, squeezing the hot metal into the desired shape. Such hammers weigh from 200 pounds (90 kilograms) to 50 tons (45 tonnes).

Many everyday articles are drop forged. One of the most common is the connecting rod for internal combustion engines in automobiles. The connecting rod is placed under a large stress when the engine is running. Forging enables the connecting rod to withstand huge stresses.

Pressing metal

Another common method of forging uses enormous hydraulic presses instead of hammers. The presses contain molds called dies. The metal is placed between the dies, the machine closes, and the metal is formed into the desired shape. One advantage of press forging is that relatively little force is absorbed by the machine and its foundations. Press forges can now shape huge metal blocks up to 250 tons (226 tonnes) in weight.

Presses come in many different shapes and sizes. Presses used to make automobile parts, for example, can be the same size as a truck. They have to be put together section by section using cranes and forklift trucks. They are often housed in separate press rooms and held together by bolts

▼ *A metalworker pours molten metal into a mold. Metals are melted and cast in factories called foundries. Casting is one of the most popular ways of converting molten metals into useful sections.*

that attach the press to the floor. The pressing action of these huge machines can generate forces of nearly 50,000 tons (45,350 tonnes).

Modern presses can often do several jobs at once. The type of press that makes the doors for automobiles, for example, can blank the metal (punch it out), shape it, and make a hole for the handle all in one stroke. They are often also part of an automated system. Robotic arms remove the blank from the press and drop it onto a conveyor belt, which takes it to the next manufacturing stage.

Rolling metal

There are various ways of shaping metal, but rolling is one of the most important. The block of metal is placed between two rollers that compress it into a thin sheet. The sheet can then be passed between more rollers that squeeze it to the desired thickness.

Rolled metal can be either hot or cold, although some parts of the rolling process must be carried out on heated metal, because hot metal is more easily flattened and shaped. Even with hot metal, however, one pass through a set of rollers only tends to compress the metal by up to 30 percent. If the metal needs to be thinner, it must pass through several sets of rollers. One set of rollers is called a stand. Rolling mills with only one stand are common, but quite often several pairs of rollers are linked together in a mill train. A tandem train has two or more stands following each other in a

▶ *Sheet steel emerges from a rolling mill at a speed approaching 60 miles (100 kilometers) per hour. At high temperatures, steel is soft enough to roll into coils for storage or further processing.*

straight line. A looping train has several stands side by side, linked by other stands. A modern rolling mill can be large and complex, with accurate rollers mounted in vast steel housings.

How rollers shape the metal

If the rollers are perfectly smooth, they produce a flat metal sheet. As the gap between each pair of rollers is made smaller, the metal is eventually compressed to the desired thickness. It can then be cut into plates or strips. These products vary from plates as thick as 1 foot (30 centimeters) for use in heavy engineering to strip metal of less than $\frac{1}{10,000}$ of an inch (a few microns) thick for the electronics

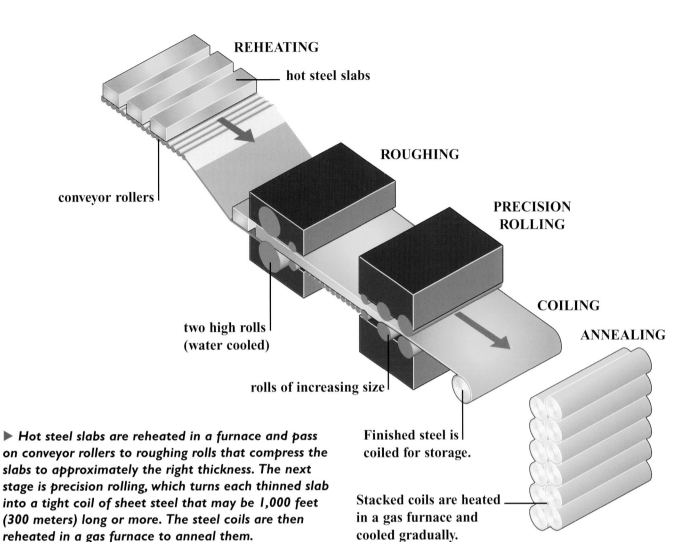

REHEATING

hot steel slabs

conveyor rollers

ROUGHING

PRECISION
ROLLING

two high rolls
(water cooled)

COILING

ANNEALING

rolls of increasing size

Finished steel is
coiled for storage.

▶ *Hot steel slabs are reheated in a furnace and pass on conveyor rollers to roughing rolls that compress the slabs to approximately the right thickness. The next stage is precision rolling, which turns each thinned slab into a tight coil of sheet steel that may be 1,000 feet (300 meters) long or more. The steel coils are then reheated in a gas furnace to anneal them.*

Stacked coils are heated
in a gas furnace and
cooled gradually.

industry. Rolling mills can also make complex shapes called sections. Passing the metal through rollers that have grooves and channels in them creates products such as bars, channels, rails, and rods. Unlike rolling flat sheets, the distance between two rollers making sections is kept the same. The final shape is achieved by passing the metal through smaller and smaller grooves. As the metal emerges from the final rolling mill, it has to be cut or rolled up. The sheet can be traveling at a speed of up to 60 miles (100 kilometers) per hour, so the shears (used for cutting) and coilers (for rolling up the metal) have to be very quick, accurate, and strong.

Smoothing and polishing

The surface of a metal object may need to be particularly hard or smooth to serve its purpose. For example, bearing surfaces on machine parts usually have to be engineered with great precision. Many metal items must also look good to attract customers. Methods used for smoothing metals include grinding and wire brushing. Another method is tumbling, in which the metal items are placed in a rotating drum similar to a tumble dryer for clothes, except that the machine is lined with small, star-shaped pieces of cast iron. The barrel spins at about 10 to 15 revolutions per minute, and after 30 minutes or so the metal surfaces are brightly polished. Electropolishing imparts a high-quality surface finish to metals. The reverse of electropolishing is electroplating, which uses an electrical current to add a layer to the surface of a metal object. One advantage of electroplating is that it can provide an attractive finish to objects with irregular surfaces—jewelry, for example, or automobile trims.

Another finishing process, called superfinishing, can be used for machine parts that rub against another surface. A lubricated (oiled) stone rubs away at the surface very slowly and gently.

Coating the finished item

Metal surfaces can be coated with enamels, lacquers, paints, and varnishes as part of the finish. Color or protection against corrosion (attack by rust or chemicals) can also be achieved by electrical and chemical means. For example, a light metal, such as aluminum, is often anodized. In this electrochemical process, a thin, dense oxide film is allowed to grow on the surface of the metal. The oxide film acts as a barrier and prevents corrosion. Steel is sometimes protected with a thin layer of zinc in a process called galvanization. Paint can also protect against corrosion. Worn metal surfaces can be coated with metals so that they can be machined to the original size once again. Electroplating is one method. Various forms of metal spraying are also used to coat the surfaces of worn metals.

DID YOU KNOW?

There are more than one million robots in factories across the world. Robots work with incredible precision and repeatability that is far superior to a human operator. High-precision robots machine metal to accuracies of one-thousandth of an inch (several microns) in a fraction of a second.

Safety and maintenance

High standards of safety are very important in metalworking factories. For example, large presses are often designed not to work unless the operator has both hands on the control buttons. In this way, the operator cannot crush his or her hands while using the machine. Another safety measure for smaller machines is to attach the operator's gloves to part of the device with a cable. If the operator accidentally puts his or her hands inside the machine, they are jerked out by the movement of

◀ *A metalworker uses a grinding machine to smooth the surface of a metal machine part.*

▶ *A quality control inspector measures the cylinder bores of an engine with the help of a machine.*

the mechanism. Some automated machines also have safety guards that must be engaged before they will operate. Repairs can be expensive, so the parts must be kept in good working order. A fraction of an inch error in placing the metal could damage the machine. Therefore it must be properly set to ensure that everything fits together properly. Lubrication (oiling) or painting with soap solution also ensures that the device runs smoothly and its parts do not wear away.

Mechanization and automation

Ancient metalworkers relied on human strength and skill to hammer metals to the correct shape. From the time of the Industrial Revolution, machines were used to bend, crush, press, and roll metals. Huge factories manufactured metal items quickly and in bulk. In turn, these products drove forward the age of industrialization, because they provided the parts for more machinery.

Further advances in technology during the twentieth century led to machines replacing people for many metalworking tasks. Automated robots and production lines can cut, press, and assemble items without direct human control. One of the

most automated industries is the design and manufacture of automobiles. In fact, engineers first used the term *automation* at the Ford Motor Company in the 1940s. Automated production lines are central not only to making the individual components but also to assembling the parts.

Today, robotic metalworkers can cut precise shapes in sheet metal with lasers and weld the parts together. Other robots spray paint or polish the finished product. The production of sheet metal in rolling mills relies on computer technology. Sensors measure the size and temperature of the metal billet and then feed data to control programs that decide how to position the rollers.

Many modern items are now too complex to manufacture by hand. Computer chips, DVD players, and cell phones are highly intricate and complex devices. In the future, scientists imagine fully automated factories with little or no human intervention. Products could be ordered over the Internet, with sophisticated robots then fabricating the products to the customer's specifications.

See also: METAL • METALLURGY • ROBOTICS

Metamorphosis

A kitten and a puppy resemble a tiny cat and dog, but a caterpillar and a tadpole look nothing like a butterfly or a frog. The process that brings about such a total change in the form and structure of certain animals is called metamorphosis.

The term *metamorphosis* comes from the Greek word that means "to transform." It is a very good description for the complete change that takes place in some animals from the young to the adult. Metamorphosis happens in the lower forms of animals, such as amphibians (organisms that live both on land and in water), insects, snails, and shellfish such as crabs and lobsters.

This complete change in shape means a complete change in lifestyle. For amphibians, metamorphosis means changing from a life in water to living on land. Frogs begin life as fishlike tadpoles in water, but they end up as creatures that jump on land as

well. These changes in lifestyle are a great advantage for small animals, because it means the young do not compete with adults for food or living space.

Lifestyle changes are usually timed to coincide with the availability of food. Caterpillars hatch from their eggs at the time when their plant food are growing in spring, then they transform into adult butterflies in time for the summer flowers on which the adults feed.

The changes are made partly in response to environmental factors, such as a change in the weather or the hours of daylight, and partly in response to chemical messengers, called hormones, in the animal's body.

It is thought that hormones called molting hormone and juvenile hormones, which are not species specific, regulate metamorphosis, flooding through the animal's body to stimulate the body to change. These physical changes, as well as those

▼ *The axolotl is a salamander that lives in Mexican lakes. What makes it strange is that it goes through egg and larval stages like other amphibians, but never metamorphoses to an adult. This is called neoteny.*

involving differentiation and growth, are accompanied by dramatic alterations of the organism's biochemistry, behavior, and physiology.

Different kinds of metamorphosis

Biologists talk about complete and incomplete metamorphosis. They call it complete if there is no resemblance at all between the young and the adult, such as in butterflies and moths. They call it incomplete if some likeness remains, such as in crabs, grasshoppers, and lobsters.

Biologists studying insects actually use three particular words to identify three different kinds of life change. When insects undergo complete metamorphosis—such as beetles, butterflies and moths, wasps, and flies—biologists call the change "holometabolous." When insects go through incomplete metamorphosis—such as bugs, grasshoppers, and termites—biologists describe it as "hemimetabolous." And when insects simply grow, without changing shape very much at all, biologists describe it as "ametabolous." This happens to insects that never grow wings, such as silverfish and springtails.

Complete metamorphosis

There are four distinct stages in the complete metamorphosis of an insect: egg, larva, pupa, and adult or imago. The stages are most dramatic in butterflies and moths, but wasps, flies, beetles, and other insects go through the same stages. But while a butterfly becomes a caterpillar in the second, larval stage of life, a beetle or a wasp is called a grub or a mealworm, and a fly is a maggot. Dobsonfly larvae are called hellgrammites. None looks remotely like the adult insect.

With all insects, the first stage of life is the egg. Each insect has its own favorite place to lay its eggs. Butterflies typically lay their eggs on the leaves of the plant that will provide food for their young when they hatch. Flies often lay their eggs in rotting meat, or in dung, on which the maggots can feed. Many beetles lay their eggs in rotting wood. When the eggs hatch, typically after just a few days, the insect enters its second stage of life.

▲ *This adult blue morpho butterfly is just emerging from its chrysalis after metamorphosing from a caterpillar. It will hang for a moment while its wings dry, then will fly off to begin its brief adult life.*

The larval stage

The second stage of complete metamorphosis is the larval stage. Larvae (*singular*, larva) are what emerge from the eggs. Larvae such as grubs and maggots are dull gray or brown, but many larvae, such as caterpillars, can be striking colors—brown, yellow, green or striped—and may have smooth, hairy, or spiny skin, depending on the kind of butterfly or moth they are going to turn into. Caterpillars move either by making a looping motion or by crawling on leglike parts called prolegs. They do little more than eat all the time, and so they have powerful jaws for chewing and a large intestine for digesting the food.

◄ *Bees are among the many insects that go through the four stages of complete metamorphosis. In this photograph, adult bees are looking after the white, grub-like larvae curled up snugly in what are called brood cells in the bees' nest. The pupal stage is spent in the brood cells, too, and the bee only emerges when it is finally an adult.*

Many caterpillars can eat several times their own weight in a single day. In fact, the larva eats so much that it outgrows its outer skin four or five times. Whenever the skin becomes too small, the caterpillar will shed it. This process, called molting or ecdysis, is triggered when hormones from the caterpillar's brain stimulate glands in its body to release a hormone called ecdysone. Ecdysone tells the caterpillar's body to grow a new, soft skin under the old one, which it is ready to shed.

Each of these growth stages of the larvae is called an instar. A caterpillar just out of its egg is in its first instar; its second instar begins after its first molt, and so on. Many caterpillars change dramatically between instars. Some, for example, resemble bird droppings or slugs in their first instars, then look more like snakes in the last instars. This change helps them survive. While they are small, they are more likely to go unnoticed by enemies if they are drab. But when they are larger, they may be so visible that it is safer to look frightening.

The pupa

The pupal stage is the third stage of an insect's complete metamorphosis. The change in this stage is so dramatic that it takes quite a long time, and the pupa (*plural*, pupae) can neither move nor feed while it is taking place. The pupal stage is often called a resting stage.

Sometimes the change happens in just a week or so. Sometimes it can last many months. A few butterflies go right through the winter as pupae, emerging only the following spring when there is plenty of food around.

While it is changing, the pupa is completely immobile, so the insect needs protection from its enemies. The insect makes itself a tough case, called a cocoon or chrysalis (*plural*, chrysalids), from its

DID YOU KNOW?

Newts and salamanders are also tadpoles in their larval stages. Two species—the mudpuppy and the axolotl—actually remain tadpoles for their entire lives.

own silk or plant and soil debris. Many butterfly chrysalids hang under leaves from silken threads. Many beetles bury their cocoons in the soil or in old tree trunks.

Although a chrysalis looks so still that it appears lifeless, inside there is a great deal of activity as the pupa changes. Gradually, the larva's body, legs, and other organs dissolve. In their place grow wings, adult legs, and body parts.

Eventually, when the transformation is complete, the pupa emerges from its skin into the fourth and final stage, when the insect is an adult or imago. Often it will not be ready to fly off immediately and must wait until its wings dry out. Adults look and behave in extremely different ways from larvae. They have wings, sometimes with beautiful colors and markings, so they fly rather than crawl. Adults may feed very little and, in some cases, not at all. They live very briefly, and only to reproduce. Once they have reproduced, they die. But the eggs they lay begin the cycle of complete metamorphosis all over again.

Incomplete metamorphosis

In incomplete metamorphosis, the third (pupal) stage does not happen, and the larval stage is called a nymph. So there are three life stages: egg, nymph, and adult. Bugs, cockroaches, dragonflies, grasshoppers, and stick insects are some of the insects that go through these stages.

The larva or nymph already has rudimentary wings or wing buds when it hatches from the eggs, and it looks quite similar to the adult. In fact, nymphs often look like wingless, miniature versions of the adults. Often, the nymph may even live in much the same kind of places and eat the same kind of food. Dragonfly and mayfly nymphs live entirely in water, though, only emerging into the air as adults. As the nymph grows, its wings gradually develop on the outside of its body, and they get larger at each molt. The number of nymphal stages or instars varies from insect to insect. Typically, there are four to eight stages.

Amphibian metamorphosis

Amphibian metamorphosis has differences from insect metamorphosis. In fact, each amphibian has its own particular way of developing. The frog is a good example.

In the first stage, the frog's eggs, called spawn, are laid in water. Each egg contains an embryo protected by a ball of jelly. After a week, the eggs hatch as tadpoles, the second, or larval, stage of the frog's life. The newly hatched tadpole has gills for breathing in water like a fish, and it uses its mouth to help it cling to underwater plants. It has a tail and looks rather like a tiny fish. Very quickly, however, the tadpole begins to grow and change.

After three weeks or so, its tail has grown long enough to help it swim rapidly through the water. It still seems fishlike, but soon it begins to grow hind legs, then lungs for breathing in air, and it begins to feed on insects. Its body begins to change shape and to resemble a frog. Not long after, the tadpole grows front legs, and it absorbs its gills into its body, so that it has to swim up to the surface to gulp for air. Finally, a tiny frog with a stump of a tail emerges from the water.

▼ *Here you see the dramatic changes in a frog's body shape, as it metamorphoses rapidly from a one-week-old tadpole to a 14-week-old adult frog.*

See also: BIOLOGY • CELL • EVOLUTION • REPRODUCTIVE SYSTEM

Meteorology

Meteorology is the science that studies the atmosphere, including weather conditions and their causes. Using this knowledge, meteorologists try to predict (forecast) how the weather will change in a given area.

The invention of scientific instruments such as the wind vane and thermometer in the sixteenth century, and the barometer in the seventeenth century, provided the first ways to systematically measure weather conditions. The modern meteorologist's essential tools still include thermometers and barometers, as well as other instruments. Much of this equipment is housed in simple weather stations. These are white, wooden boxes that stand about 4 feet (1.3 meters) above the ground. The sides of the boxes are slatted so that air can flow freely through

them, while the top of the boxes shade the thermometers inside. Some simple thermometers are used to measure air temperatures, but most weather stations contain a hygrometer. This measures not only air temperature, but also humidity (the water content of air), using a wet-and-dry-bulb thermometer.

The dry-bulb thermometer records the actual air temperature. The wet-bulb thermometer has a piece of muslin around its bulb that is kept wet. Moisture evaporates from the muslin, and this lowers the temperature on the wet bulb thermometer. Since the rate of evaporation depends on the humidity of the air, the relative humidity of the air is calculated from the difference between the readings on the wet and dry thermometers.

Air pressure is measured with barometers, and wind directions and speeds with wind vanes and instruments called anemometers. Rain gauges—containers used to collect rainfall—are sunk into the ground. This reduces loss of moisture by

▶ *This image of a large hurricane was taken by a weather satellite. Observations of weather systems from satellites have been a significant help to meteorologists in their understanding and prediction of Earth's weather patterns.*

◀ *A standard barometer and thermometer in a simple home weather gauge. Despite their simplicity, barometers and thermometers are still important tools of the meteorologist.*

evaporation. Sunshine recorders contain a glass sphere that focuses the Sun's rays onto a card. As the Sun moves across the sky, a line is burned across the card. Meteorologists also record other features of the weather, such as cloud types, cloud heights, visibility, and so on.

The evolution of meteorology and forecasting

Advances in scientific knowledge during the seventeenth and eighteenth centuries established the physical basis for modern meteorology, helping scientists to understand weather. However, it was not possible to forecast weather until 1844, when U.S. inventor Samuel F. B. Morse (1791–1872) perfected a wire communication system called the electric telegraph. With the electric telegraph it became possible to collect information on weather conditions from many different weather stations quickly enough to prepare weather forecasts. Attempts to forecast weather had begun by the late 1840s, and soon newspapers were publishing weather reports. In 1870, a national weather service was set up in the United States as part of the Army

Signal Corps. In 1890, the civilian Weather Bureau succeeded this service. It was renamed the National Weather Service in 1970.

The study of weather systems developed in the late nineteenth century. The connection between depressions (also called cyclones) and warm and cold air masses was made by British Admiral Robert Fitzroy (1805–1865). Fitzroy also began a system of issuing storm warnings. The greatest advances were made in Norway during and after World War I (1914–1918). A group of meteorologists, led by Dutch professor Vilhelm Bjerknes (1862–1951), put forward the Bergen theory to explain how depressions form. They identified cold fronts, warm fronts, and occlusions (created when a cold front overtakes and undercuts a warm front). The understanding of the importance of warm and cold air masses, and the way they interact along fronts, led to a great advance in the accuracy of weather forecasting.

Before World War II (1939–1945), meteorologists could only study the behavior of the weather near ground level, although they did get some

information from airplane and balloon pilots. During the war, however, radiosondes came into widespread use. Radiosondes are hydrogen-filled balloons with self-recording instruments to measure air pressure, humidity, and temperature higher up in the atmosphere. Radiosondes also contain radio transmitters that relay measurements back to the ground. As the balloon soars upward, a target attached to the balloon is tracked by radar, enabling meteorologists to figure out the wind speeds and directions in the atmosphere. Every day, hundreds of radiosondes are released from weather stations all over the world.

Weather satellites, which have orbited Earth since 1960, are another source of information. They send back pictures of cloud formations, which help forecasters prepare weather maps. They also record the movements of storms. Weather satellites also measure heat and record information about conditions at various levels of the atmosphere.

▲ *Meteorologists supply weather information to television stations to broadcast weather forecasts. Although computers are now used in forecasting, the skill and experience of meteorologists is still important.*

Information about the upper atmosphere has increased the understanding of weather systems, such as depressions (lows) and anticyclones (highs). It has also shown the existence of powerful high-altitude winds, called jet streams.

Radar is also used to gather information about precipitation (rain, sleet, hail, and snow). During World War II, radar operators noticed that airplanes escaped detection when they flew into rain clouds. This was because the radar beam was stopped by the ice crystals and water droplets in the clouds. This discovery has been developed so that special kinds of radar are now available for use at weather stations to locate where snow and rain are falling. Radar can detect falling rain more than 200 miles (320 kilometers) away.

▶ *A radiosonde (weather balloon) is released from a weather station. Radiosondes provide detailed information of weather conditions at high altitudes.*

When all the measurements have been made, the meteorologists put the data into code. This coded information is then relayed to weather centers. The code used is an international one so that it can be understood anywhere.

Information about weather conditions is collected at weather stations. These stations are now located all over the world, both on land and at sea. The U.S. National Weather Service has about 300 weather stations in the United States and its overseas territories. There are also more than 12,000 substations. At weather stations, the meteorologists measure the conditions of the air every three or six hours. At a few stations, data is gathered every half hour.

International cooperation is vital in modern weather forecasting. In many countries, the weather systems on one day developed a day or so before in another distant country. For example, the National Weather Service exchanges weather information with both the Canadian Meteorological Service and the World Meteorological Organization. The World Meteorological Organization is an agency of the United Nations. Its meteorologists collate the data from its more than 150 member countries every 12 hours. For smaller areas, it provides information every three hours, or sometimes even more often.

In recent years, the chief developments in the field of weather forecasting have been the use of computers, radar, radiosondes, and weather satellites. The increasingly detailed study of the changing weather patterns on Earth has made more accurate forecasts possible.

Meteorological research
Meteorologists are always studying the atmosphere to better understand atmospheric conditions and what causes certain types of weather. This has helped meteorologists to make more accurate forecasts, to give early warning of storms, such as blizzards, cyclones, hurricanes, and tornadoes, and even to learn how effectively snowflakes clean pollutants from the air.

DID YOU KNOW?

There are about 10,000 weather stations around the world, on land or on ships in the open sea. At most of these stations, meteorologists record measurements of weather conditions at regular intervals throughout the day and night. Some weather stations in remote areas have no regular staff, so the measurements are recorded automatically.

A major focus of meteorology in recent years has been the buildup of methane (CH_4), carbon monoxide (CO), and, to a lesser extent, carbon dioxide (CO_2) in the atmosphere. These gases trap heat rising from Earth's surface, warming the atmosphere and creating what is called the greenhouse effect. Over a long period, this greenhouse effect would warm up Earth's climate, and scientists are concerned about the possible problems this could cause.

Is the climate changing?

Climate is the long-term weather of a place, and there are fears that human activities may be changing the world's climate. Climatologists study the world's weather to determine if, for example, it is getting colder or warmer, drier or wetter.

If, as it appears, the world's climate is warming, sea levels will rise. The appearance of holes in the atmosphere's protective ozone layer also means that new agricultural methods may be needed. It is the task of climatologists to measure and predict such changes. Horticulturalists will need time to develop new varieties of plants, and farmers will have to change their farming methods.

Scientific bodies around the world have set up the Global Climate Observing System to discover how the climate is changing. In fact, some human activities have almost immediate effects on the climate. When a forest is cut down, for example, a greater amount of sunlight is reflected back from the ground. The surface heats less, and there is less rainfall. In turn, this means that the land becomes drier. This certainly happened over large areas of the Near East and Middle East in prehistoric times. The climate change may even have led to the decline of some early civilizations.

One dramatic finding from studies of past climates is that the climate can change very quickly. There are some gradual changes, taking centuries to occur. But ice ages can begin and end very rapidly,

▼ *By understanding local climate and predicting changes, humans can plan to cope with disasters such as the flooding of the Tauber and Maine Rivers in Germany.*

▶ *Meteorologists at a hurricane center in Miami, Florida, monitor weather information and try to predict where and when hurricanes might develop.*

maybe over no more than 50 years. They are caused by slight changes in Earth's orbit around the Sun and shifts in the pattern of continents on the Earth's surface.

Between about 1550 and 1850 there was a "little ice age." Average temperatures were about 2 to 5°F (1 to 3°C) colder than they are now. The traditional winter snow scenes were a common sight then. These days, widespread snow is becoming rare in Europe. Climatologists believe that if it were not for human activities, there would be a return to cold winters in the next century. For the last 10,500 years, since the last ice age ended, Earth has been enjoying a warm period. But the signs are that, in the next few thousand years, frozen conditions could return.

The World Ocean Circulation Experiment, which began in 1990, set out to monitor the ways in which the world's ocean and air currents circulate. Much of this was done using satellite observations. In the early 1990s, for example, average wave sizes in the North Atlantic were 50 percent higher than during the 1960s. In March 1993, the eastern seaboard of the United States was devastated by the worst storms in memory. As the wrecked oil tanker *Braer* leaked its cargo into the sea off Shetland, Scotland, in January 1993, the atmospheric pressure was the lowest since records began a century ago. The storms were the fiercest that had been known.

Are the apparent changes that are seen around the world really permanent? Or are they just temporary variations that people need not worry about? The longer meteorological records extend, the more extremes are likely to be seen. There are bound to be occasional long runs of hot, dry years or cool, wet years. Long-term records will help people to decide what is normal.

Computers

Large, modern computers can do the work of thousands of mathematicians. For meteorologists, and particularly for climatologists, the way to study what might happen in the future is to construct a computer model of the climate from mathematical equations. One important process that has been modeled in this way is the El Niño Southern Oscillation (ENSO), which is the fluctuation of ocean temperatures in the eastern and central Pacific Ocean. These changes can have huge climatic implications around the world.

Computers are now installed in forecast centers such as the U.S. National Weather Bureau. Some forecasts are now done by computer, especially for conditions higher in the atmosphere. But computer forecasts of weather conditions at ground level are still unsatisfactory, because computers cannot allow for all the complex interactions between land, sea, and air. Most of the weather forecasts on radio and television are, therefore, still the work of skilled teams of meteorologists, who apply their own judgment to the mass of detailed information they receive.

See also: AIR • BAROMETER • CLIMATE • CLOUD • RADAR • RAIN AND RAINFALL • SATELLITE • THERMOMETER • TORNADO • WEATHER SYSTEM

Microbiology

Microbiology is the study of microscopic life, such as algae, bacteria, fungi, protists, and yeasts, and also minute particles that can cause disease, such as plasmids, prions, and viruses. Studying microbes helps find ways to treat the diseases they cause. Microbiologists also look at microbes involved in food production and how microbes can be used in industrial processes. They are also involved in the rapidly developing technology of genetic engineering.

◀ French scientist Louis Pasteur not only proposed the germ theory of disease but also showed how foods could be heat-treated to destroy microbes that spoil them. In honor of his work, the treatment is now called pasteurization.

Until the seventeenth century, no one suspected that there were aspects of life too tiny to be seen by the naked eye. Around 1660, English scientist Robert Hooke (1635–1703) started to use the newly invented microscope to study plants. In his remarkable book *Micrographia,* he described tiny sections of leaves and drew what he saw, which were minute boxlike structures now called cells. Later, scientists discovered that all living organisms consist of cells. The study of cells is now the cornerstone of microbiology.

Meanwhile, on continental Europe, scientists such as Dutchman Anton Leeuwenhoek (1632–1723) were using microscopes to study the world in detail. Leeuwenhoek made an astonishing discovery when he looked at a drop of pond water through a simple microscope. He saw a world of organisms far too small to see with the naked eye. Leeuwenhoek called them "animalicules," but scientists now call these tiny creatures protists. He also observed bacteria in decaying organic matter.

Many more scientists began to study bacteria after Leeuwenhoek, but few people suspected their role in causing disease. In the 1870s, dramatic improvements in the power of the microscope allowed German scientist Ferdinand Cohn (1828–1898) to classify bacteria, thus founding the science of bacteriology. Then French scientist Louis Pasteur (1822–1895) and German physician Robert Koch (1843–1910) developed the germ theory of disease, showing that microbes like bacteria were responsible for many diseases. The importance of studying microbes immediately became apparent.

Soon, some scientists began to realize there were even smaller disease-causing particles—in fact, too small to be seen even with a microscope. These tiny particles, now called viruses, were first discovered in tobacco leaves by German agricultural chemist Adolf Mayer (1843–1942). Dutch bacteriologist Martinus Beijerinck (1851–1931) independently discovered the same virus in 1898.

Discovering DNA

One more major discovery was needed to lay the foundations of modern microbiology—unlocking the secrets of deoxyribonucleic acid (DNA). DNA is found inside most living cells, and it is the genetic blueprint for life. DNA was first discovered in 1869 by Swiss physician Johann Friedrich Miescher (1844–1895), but he had no idea of its importance.

Not until the 1950s did scientists begin to appreciate DNA's role in biology. In 1952, U.S. biologist James D. Watson (1928–) and English scientist Francis Crick (1916–2004) realized that a DNA molecule has a double helix shape—similar to a rope ladder twisted in a spiral. After this discovery, microbiologists began to look at the structure of DNA using powerful electron microscopes and then put it together again to find out how it passes on genetic information. In 1967, the mechanism underlying DNA, called the genetic code, was finally unraveled by U.S. biochemist Marshall Warren Nirenberg (1927–) and Indian-born U.S. biochemist Har Gobind Khorana (1922–). Soon a new area of microbiology began to develop to study and manipulate DNA.

Microscopes

A key factor in the progress of microbiology has been the development of microscopes. Even the most powerful light microscopes—microscopes using just lenses to magnify the image—can only magnify up to 2,000 times. But in the 1960s, microbiologists began to use electron microscopes, which can magnify up to one million times. With electron microscopes, microbiologists could see not only viruses but also tiny infectious particles, such as virions, which are small parts of the protein coat of a virus. In the 1980s, even more powerful microscopes appeared, including the scanning tunnelling microscope (STM), which allowed microbiologists to study microbes and living cells right down to the atomic level.

▶ *Agar is a red jelly extracted from seaweed. It is the perfect medium (food) for growing cultures of bacteria for laboratory work.*

Microbiology in action

The key role that microbes play in disease has always made microbiology much more than just an academic subject. At the same time that academic microbiologists try to understand diseases, others have been looking for ways to apply this knowledge to find cures and preventive treatments, including antibiotics and vaccines. Microbiologists have also explored ways to apply their knowledge to everyday life, aiming to find out how microbes in soil affect fertility, how they affect the health of water supplies, how they spoil food and make it safe or unsafe to eat, and how they are needed to make products such as cheese, wine, and yogurt. In recent years, the study of genetic engineering (genetic modification) has taken applied microbiology in an entirely new direction.

Bacteria

Bacteria are about $\frac{1}{200,000}$ inch (0.0005 millimeters) across. They may be rod-shaped, spherical, comma-shaped, or coiled. They belong to a class of organisms called prokaryotes, which means they have no nucleus. Their DNA is in their cytoplasm.

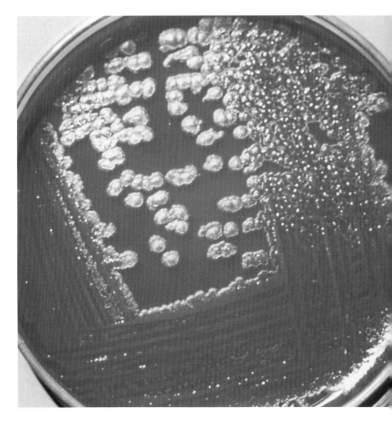

Bacteria grow rapidly. Under ideal conditions, the bacterium *Escherichia coli,* which is found in the human intestines, will divide every 20 minutes. Since they multiply so rapidly, bacteria are widely used in research and industrial applications. Genes from other organisms have been inserted into *E. coli* to make them create products such as human growth hormone (HGH) and insulin. To make these products commercially, bacteria are grown in huge vats called fermenters, holding up to 25,000 gallons (100,000 liters).

Some bacteria manufacture antibiotics, such as streptomycin and tetracycline. No one knows why bacteria should make substances that kill bacteria—it may be a defense against competition. Some bacteria from the genus *Lactobacillus* help make processed dairy foods such as yogurt. Bacteria may even find future applications in biological control of pests. One already in use is *Bacillus thuringiensis,* which secretes an insecticide.

▲ *Called noble rot, the fungus* Botrytis cinerea *looks unpleasant on a strawberry. However, winemakers use it because it shrivels grapes and intensifies their flavor.*

Fungi

Many fungi also secrete antibiotics, especially species from the genus called *Streptomyces.* But penicillin, from a mold called *Penicillium notatum,* is the best-known example. Common fungi are shaped like threads and grow from the tip. Mold on stale bread is actually colonies of fungi. Under a microscope, one can see new growth occurring around each colony's outer edge. Studying fungi may yield new antibiotics. Several species of fungi are also used in cheese manufacture.

Fungi called yeasts have many commercial uses. They are essential in the baking and brewing industries. Similar to bacteria, yeasts are single-celled organisms, but they have a much more complicated cell structure. Yeasts are therefore more like higher organisms than bacteria, so it is easier to insert genes from higher organisms into yeasts. Yeasts may turn out to be even more useful than bacteria for biomanufacturing.

Viruses

Viruses are much smaller than bacteria and consist simply of DNA or ribonucleic acid (RNA) surrounded by a protein coat called a capsid. The DNA or RNA carries genetic data for making more copies of the virus, but the virus lacks the machinery for doing so. To reproduce, the virus must enter a cell and "hijack" the cell's systems for making protein and DNA. The big drive in virology (the study of viruses) is trying to understand how viruses infect and cause disease in humans. The human immunodeficiency virus (HIV, which causes acquired immunodeficiency syndrome, or AIDS) is the target of intense research.

Clues to the origins of life

Many bacteria are very resistant to heat, cold, high salinity (salt concentration), and desiccation (drying out). As a result, many scientists believe bacteria probably played an important part in the early development of life on Earth. Some theories concerning the origins of life suggest that the first organisms to appear, some time around one billion years after Earth's formation, were simple bacteria. For more than two billion years after that, microorganisms were the only life on Earth. One kind of bacteria, the blue-green algae, developed photosynthesis—the process by which energy is harnessed from sunlight to make food. Then oxygen began to accumulate in Earth's atmosphere, paving the way for more complex life-forms.

Bacteria that may resemble primitive species are still found in hostile, oxygen-free environments such as those in which life probably arose. They are called archaebacteria. One place they are found is the hot springs in Yellowstone National Park. Other surprising discoveries include bacteria near hydrothermal vents deep on the ocean floor, where material emerges from below Earth's crust. More startling still was the discovery in 1992 of bacteria in drill samples taken from great depths in Earth's crust.

▲ *A microbiologist investigates* **Campylobacter,** *which is a bacterium that lives on chickens and causes 2 million cases of food poisoning every year.*

Cells of more complex organisms than bacteria contain small structures called organelles, including mitochondria and chloroplasts. Several characteristics of mitochondria and chloroplasts suggest that they might once have been free-living bacteria that were absorbed into the cells that they are now part of.

See also: BACTERIA • DISEASE • DNA • FUNGI KINGDOM • MICROSCOPE • VIRUS, BIOLOGICAL

Microelectronics

Consumers expect modern electronic gadgets to be increasingly capable. In order to make such "smart" gadgets, complex electronic circuits have to be packed into tiny spaces—a technology called microelectronics.

When a person uses a cellular phone, he or she can dial a stored number with one press of a key. The cell phone can display the name and number of a caller when a call is answered and can even take photographs and send and receive them. Automatically and unnoticed by the user, it switches its radio frequency as needed as its location changes.

A mobile phone is "smart"—which means it can do many different things with a minimum of control by the user. In the home, all the electronic household gadgets are "smart" to varying degrees—the washing machine, the stereo, the radio and television and their remote-control units, the personal computer, and the pocket calculator. All can perform a range of complicated functions, yet the electronic "brains" that do this are housed in tiny "microchips" no bigger than a fingernail.

Before the microchip

In the 1950s, when the first consumer electronics became available, radios, TVs, and other electronic devices were bulky. They contained components called valves or electron tubes. These typically looked like glass bottles a few inches long. They got very hot and they had short lives.

Tubes did all the work of controlling electric charge and electric current in these devices. They stored charge and switched and amplified currents (made them stronger). For example, they made the weak current coming from a radio antenna stronger, so that it could drive a loudspeaker and make sounds.

In the 1950s, the electron tube was increasingly replaced by the transistor and other semiconductor devices. These were made of pieces of crystalline material, housed in a small unit less than an inch across. They used much less current, did not get so hot in use, and kept working for much longer. Electronic devices that used these were smaller and lighter than tube-operated devices. Portable "transistor radios" became common, for example.

An electronic device could contain dozens, hundreds, or even thousands of these components. They had to be connected by complex networks of wires or metal strips. Usually they would be mounted on a board carrying a printed circuit. The connecting circuit was printed on this printed circuit board (PCB) in the form of connecting strips of copper or other metal. The transistors and other components would be attached to the board at the correct places.

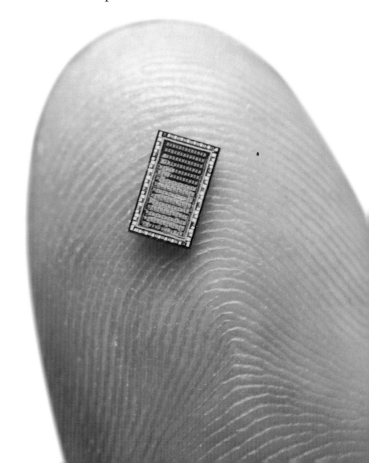

▶ *Powerful microchips can now be made incredibly small. This tiny little chip balanced on a finger can provide enough processing power for a small hand-held computer.*

▶ *A microscope photograph reveals all the minute electrical connecting strips on a printed circuit board that link into the central microprocessor.*

The next big advance was to make a group of components, forming a small circuit, as one component. This was called an integrated circuit, or IC. A number of integrated circuits could be mounted on printed-circuit boards.

The next step was very large-scale integration (VLSI). This was the process of making a circuit consisting of thousands of components in a small piece of material in one step. One such circuit was a microprocessor—the heart of a computer. Microprocessors were first used in pocket calculators and in wristwatches. Later they were used in the first personal computers.

DID YOU KNOW?

A typical microchip contains millions of transistors. These are basically electronic switches. The total number of transistors produced in microchips and other microelectronic components every year is now greater than the number of letters of the alphabet and numerals that are printed every year. And they outnumber the world's ant population by somewhere between 10 and 100 times.

Making a microchip

VLSI circuits are called microchips. Making a microchip, or VLSI circuit, begins with making a crystal of very pure semiconductor material— usually silicon. A cylinder of silicon is made in the laboratory. It is almost perfectly pure—no more than one atom in ten billion can be an "impurity" atom. All the silicon atoms are arranged in an almost perfect crystal pattern, or lattice. As many as 200 microchips will be formed at the same time in the top face of the silicon cylinder, which is typically 8 or 12 inches (20 or 30 centimeters) wide.

The cylinder is put in an oven, where atoms of oxygen enter the top face of the cylinder. This forms a layer of silicon dioxide, which is electrically resistive—it does not allow electric current to pass. Then the top face of the cylinder is coated with a material called photoresist. Ultraviolet (UV) light is shone through a mask, or transparent film. The mask contains a pattern of opaque (light-blocking) lines representing just one layer in the complex three-dimensional network that will be formed in the final microchip. Hundreds of identical copies of the circuit pattern are carried on the mask.

The UV light is focused down onto the top face of the cylinder. The ultraviolet light makes the photoresist harden. Then chemicals remove, or etch

▲ Microchips are made in a clean room. Microchips are very sensitive to any contamination, especially from moisture and human sweat, so they must be manufactured in absolutely clean conditions.

away, the unhardened parts of the photoresist, representing the circuit. In the places where the photoresist has been removed, the oxygen-containing layer of the silicon directly beneath is also removed. This uncovers the pure silicon beneath. Finally, the remaining photoresist, but not the silicon dioxide beneath, is removed with a different chemical reaction.

What is left is a complex network of pathways of pure silicon, separated by areas of non-conducting silicon dioxide. But this is still a long way from the final microchip.

So-called impurity atoms have to be added to the silicon in carefully controlled amounts, in just the right places, to form tiny components such as transistors. Then another coating of photoresist is applied, and UV light is shone through a different mask. After etching, another layer of the microchip is formed. Then more impurity atoms have to be added to form more components.

The process is repeated again and again, and coatings of metal are used as well in some places. Finally, the whole structure of the microchip has been formed in the top layer of the silicon rod. An array of hundreds of the microchips has been formed simultaneously. The last step is to slice off a wafer of the silicon cylinder and then cut this up into the individual circuits. These pieces are chips of the wafer—hence each circuit is given the name *chip* or *microchip*.

Then hundreds of fine gold wires, thinner than a human hair, are attached to the chip. These connect it to the larger metal pins that will join it to other components. The chip and its connectors are put into protective plastic packaging. Now the chips can be shipped off in their thousands to other companies who will use them in everything from computers and cars to credit cards.

DID YOU KNOW?

In 1965, George Moore, founder of Intel Corporation, now the leading microchip manufacturer, predicted that the number of components that could be packed into a given space on a microchip would double every year. Later, Moore decided that every 18 months was closer to the mark. This "law" has held good ever since. Chip manufacturers battle to invent new technologies to make sure that "Moore's Law" stays true into the future.

The microchip revolution

Microchips work faster than the full-sized circuits they replaced, use much less electrical energy, and are highly reliable. And because they are so small, they make it possible to pack enormous computing power into a tiny space.

The coming of the microchip began a revolution in electronics that is still continuing. All electronic devices were rapidly miniaturized. Personal radios and stereos and small TV receivers were made possible by microelectronics. Computers that had consisted of roomfuls of costly equipment shrank and appeared on the desktops first of business people and then in the home. Then they became portable, in the form of laptop and palmtop computers. Palmtop computers began to merge with cell phones, which became possible only when microelectronics could equip them with the powerful computing abilities needed to handle radio traffic among millions of users.

The future

One measure of the power of a chip is the number of bits that it can store. A bit is the numeral 0 or 1, and is the basic unit of information in computing. Computers are able to do all their computing and information processing representing all information as just strings of 0s and 1s. For example, the Intel Pentium 4 microchip, which was introduced for personal computers in the year 2000, can store 42 million bits. Microchips able to store a billion bits will soon be in use.

To make progress in the future, the microchip's transistors and other components must work even faster than they do today. Components work faster when they are smaller, and when there are more of them working together. As a result, the smaller the components of the chip can be made, the faster the chip will work.

▶ *Computer chips are manufactured by beaming a light through a stencil, or mask, to cast an image of the circuits onto disks of silicon called wafers. Each wafer may hold a hundred or so tiny integrated circuits or microchips.*

To make small structures in a microchip, very short-wavelength light must be shone through the mask. Today's advanced microchips have structures about 100 nanometers across, made with ultraviolet light of wavelength around 200 nanometers. (One nanometer—symbol nm—is one billionth of a meter.) Engineers are now moving from "deep-ultraviolet lithography" to "extreme-ultraviolet lithography," using wavelengths of 11–13 nm, which will be able to create structures as small as 30 nm across. These chips will be up to 100 times faster than today's most powerful chips. Computer memory chips will be able to store one hundred times as much information.

Engineers will have to find ways to reduce the amount of electrical energy consumed by these components. Although solid-state devices use far

> ▼ *Open up any piece of electronic equipment and you will see something like this—an array of complete microcomponents linked together by connecting strips on a printed circuit board.*

DID YOU KNOW?

The transistors in advanced microelectronic components can turn on and off a trillion times per second (one thousand billion times per second.) To turn a light switch on and off a trillion times would take a person more than 15,000 years.

less energy than the old tube devices, the little they do produce becomes a problem when so many components are packed into such a tiny space. Present-day computers have bulky and noisy fans to keep their microchips cool, and the problem will become worse with future computers.

Advanced nanotechnology

Still further progress will depend on advanced nanotechnology. This term refers to engineering on a scale of nanometers, when it becomes possible to

assemble things atom by atom. Some scientists even suggest this may one day be done by nanobots—robots only a few nanometers in size, assembling a microchip the way that human workers now assemble a skyscraper.

Future uses of microchips

In the future, tiny microchips, able to communicate with the outside world "wirelessly"—that is, by radio—will be implanted in more and more objects, and will be able to do more and more things. Your refrigerator will be able to tell you what's in it, and what you're running low on. You'll be able to find any schoolbook in the house—its microchip will tell you where it is. You will converse with your computer, radio and TV, rather than pressing keys. When you add new devices in your home, such as new speakers to your sound system,

they will automatically connect with the rest of the system without your having to plug anything into anything else.

In the future, microchips will increasingly be implanted in human bodies and linked to the nervous system. This has already been done experimentally. One day microchips may be able to control diseases of the nervous system, such as Parkinson's disease. Or they may bring sight to the blind by taking over the job of the retinas at the back of the eyes, converting patterns of light waves into electronic signals to pass to the brain.

Identity implants

Implanted microchips may even carry personal information. All the information carried in a passport, on credit cards, store cards, health records, academic diplomas and business cards, together with personal and work histories, could be carried in one implant.

It is already common to implant identity microchips painlessly into the body of a pet animal. The microchip contains the owner's name, address, and telephone number. If the pet is lost or stolen, it can easily be reunited with its owner.

Microchips permanently embedded in the human body could be updated continuously. Any of this information could be transferred instantly to a reading device at a customs post or in a police station or in a company that a person wished to make themselves known to. The embedded microchip could broadcast details of the location of a person to emergency services if he or she got into trouble. All these functions could be very convenient for people in their everyday lives—but civil-liberties groups are concerned that such availability of information could be misused.

◀ *Increasing miniaturization of microchips means that palmtop computers—computers you can hold in your hand—contain the computing power that only huge desktop computers had a decade ago.*

See also: COMPUTER • ELECTRIC CIRCUIT • ELECTRONICS • PRINTED CIRCUIT

Micrometer

The micrometer is an extremely accurate measuring device that is small enough to hold in the hand. Engineers use the micrometer to measure the lengths of small objects. An object is placed between the ends of two metal rods. The instrument is adjusted until the rods just touch the object, and the length measurement is shown on a scale.

Micrometers are designed to measure the thickness of small balls, rods, wires, and slots, sometimes down to one-millionth of an inch across. Some micrometers can even measure the thickness of a human hair. However, the micrometer was originally designed to measure precisely the angle between stars in the night sky. It was invented by English astronomer William Gascoigne (1612–1644) in the seventeenth century.

Gascoigne was killed during the English Civil War (1638–1660), but his instruments were rescued and brought to English scientist Richard Towneley (c. 1629–1707) at the Royal Greenwich Observatory in London. With English astronomer John Flamsteed (1646–1719), Towneley measured the position of Jupiter's moons with such accuracy that it was crucial in proving Isaac Newton's then new theory of gravitation. In the eighteenth century, scientists began to realize just how useful the micrometer could be for measuring objects other than the distances between stars in the sky, and so the modern instrument developed.

How micrometers work

The modern micrometer has a rigid, U-shaped metal frame. At one end of the U is a short, fixed metal rod called the anvil. Passing through the other end of the U is a second, longer rod called the spindle. Most of the spindle is enclosed in a metal sleeve. Part of the spindle has a fine, accurately cut screw thread. This thread allows the spindle to be screwed into or out of the sleeve.

The spindle is adjusted by means of a tube that fits over the sleeve. The protruding end of the tube is shaped to form an adjusting knob that can be gripped easily. The other end of the tube is called the thimble. As the spindle is unscrewed, the thimble moves along and uncovers a scale marked on the sleeve. The point at which the thimble cuts this scale shows the separation between the anvil and the end of the spindle.

Taking a measurement

To take a measurement, the spindle is first withdrawn so that the object can be placed against the anvil. Then the spindle is screwed in until the

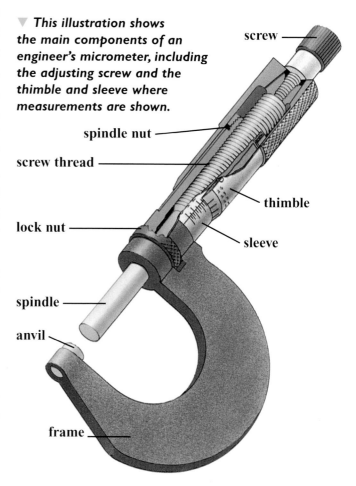

▼ This illustration shows the main components of an engineer's micrometer, including the adjusting screw and the thimble and sleeve where measurements are shown.

screw

spindle nut

screw thread

thimble

lock nut

sleeve

spindle

anvil

frame

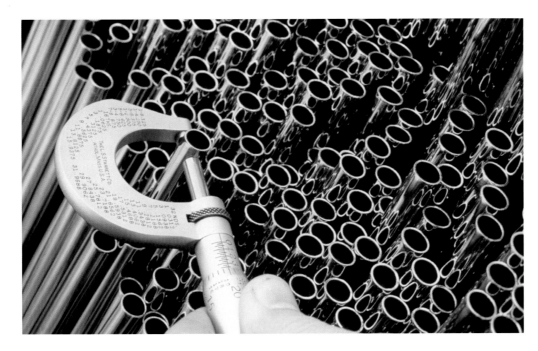

object just fits between the anvil and spindle. The reading on the sleeve scale will show the distance across the object. The spindle must not be screwed in too tightly, as this may compress the object and result in an incorrect reading.

To avoid these problems, micrometers are usually fitted with a ratchet mechanism. This links the knob and thimble to the spindle. Turning the knob screws the spindle in or out. But, when the object being measured is gripped between the anvil and spindle, the ratchet slips as the knob is turned further. This prevents the object from being gripped too tightly.

Accuracy

Although the required measurement is shown on the sleeve scale, this cannot be read with great accuracy. For example, suppose that the object in the micrometer actually measures 0.758 inch (18.95 millimeters). The scale on the sleeve would show a reading of just over ¾ inch (18.75 millimeters). But exactly how much over ¾ inch (18.75 millimeters) would not be clear.

To overcome this problem, a second scale is marked around the thimble. It shows how many thousandths of an inch to add to the reading obtained from the sleeve scale. In the example above, the scale on the thimble would give a reading

of 8 (2), meaning $\frac{1}{125}$ inch (0.2 millimeters). Adding this to the first reading of ¾ inch (18.75 millimeters) gives an accurate measurement of $\frac{379}{500}$ inch (18.95 millimeters). Some micrometers are accurate to $\frac{1}{10,000}$ inch (0.0025 millimeters).

Star micrometers

Micrometers are still used as Gascoigne intended, to measure tiny apparent distances seen in telescopes and microscopes. The micrometer used in astronomy is the filar micrometer and was developed by German-born Russian astronomer Wilhelm von Struve (1793–1864) in the nineteenth century to pinpoint double stars. The idea of the filar micrometer is to measure precisely the angular separation between two stars or other bodies in the night sky. This is, effectively, the distance between them as they appear from Earth.

The filar micrometer is put into the light path of the telescope and projects an image of the two stars in the eyepiece. The astronomer has to adjust the micrometer until the two stars appear to be in precisely the same place. The telescope is then moved to line up on each star in turn, using thin lines or "cross hairs" to calibrate its position.

See also: MEASUREMENT

Microscope

A microscope is a device that magnifies tiny objects so that they can be seen clearly by the human eye. In the four hundred years since the invention of the first microscope, these instruments have revealed a host of previously invisible worlds—from the mass of microscopic creatures that inhabit the world to the atoms that make up every substance in the universe.

No one knows who made the first microscope. Craftspeople probably used drops of water as magnifying glasses thousands of years ago, and the Romans may have made magnifying glasses with rock crystal. However, it was Dutch eyeglass maker Zacharias Janssen (1580–1638) who is credited with the invention of the first microscope. Janssen was skilled in grinding glass. His breakthrough was to put two glass lenses together to dramatically increase the magnification.

A carefully crafted magnifying glass can magnify things at best up to 20 times. However, a double-lens microscope, called a compound microscope, can magnify objects many hundreds or even thousands of times. One lens is held close to the object and is called the objective lens. It provides the initial magnification by bending light rays apart to enlarge the object being viewed. The other lens is held close to the eye and magnifies the image in the object lens. It is called the eyepiece, or ocular.

Scientists soon realized the importance of Janssen's compound microscope, and they began to make their own improvements.

In 1665, English scientist Robert Hooke (1635–1703) published a book, entitled *Micrographia*, in which he explained the basic principles of microscopy—and also made the first drawing of a minute living cell that he had observed in a slice of cork. Janssen's microscope had just been a hand-held tube, like a simple telescope today. Hooke mounted his microscope on a stand to keep it steady enough to observe every tiny object. This design soon became standard for all microscopes.

Discovering microorganisms

Meanwhile, Dutch naturalist Antoni van Leeuwenhoek (1632–1723) was making the most astonishing discovery with his own microscope. Leeuwenhoek's handheld microscope had only one lens, but he ground his tiny lenses so perfectly that they could magnify by more than 200 times.

Through them, Leeuwenhoek observed a miniature world of tiny creatures in drops of pond water. He called them animalcules, but scientists now know that these were the first observations of microorganisms such as bacteria and protists. The discovery was so astonishing that few people believed him at first.

Soon, other scientists were making their own amazing discoveries with the microscope. Italian anatomist Marcello Malpighi (1628–1694) observed tiny blood vessels, called capillaries, that link the inflowing and outflowing halves of the body's blood circulation system. Dutch naturalist Jan Swammerdam (1637–1680) discovered red blood cells. English botanist Nehemiah Grew (1641–1712) observed the microscopic difference between female and male plants.

◀ This is an early compound microscope similar to the one made by Janssen in the 1590s. Basically, it is a tube with a lens at each end, one for magnifying the object and the other for enlarging the magnified image.

The perfect lens

At this time, microscopic lenses suffered from the same two faults as the early telescopic lenses. One problem was that the image they gave could not be made sharp all over, only at the center or the edge. In addition, the image viewed using a microscope was surrounded by a fringe of color. In the 1830s, English wine merchant and scientist J. J. Lister (1786–1869)—father of English physician Joseph Lister (1827–1912)—solved the two problems of clear focus and color interference.

During the nineteenth century, microscopes became a vital part of every scientific laboratory. At least three manufacturers in England, two in other parts of Europe, and one in the United States became famous for the high quality of the instruments they produced. One of these well-known manufacturers was German businessman Carl Zeiss (1816–1888), who went into partnership with German physicist Ernst Abbe (1840–1905). Abbe studied the theory of the microscope in detail and succeeded in making important advances in lens design. By 1870, all Zeiss lenses contained the improvements discovered by Abbe.

By 1882, most glass for microscopic lenses was being made from a formula created by German chemist and optician Otto Schott (1851–1935). Schott's lenses were the most perfectly corrected

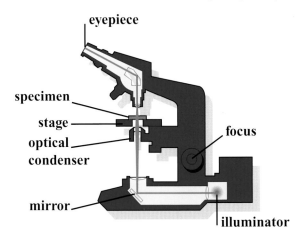

▲ A basic optical microscope lights the specimen from underneath, often reflecting it off a mirror.

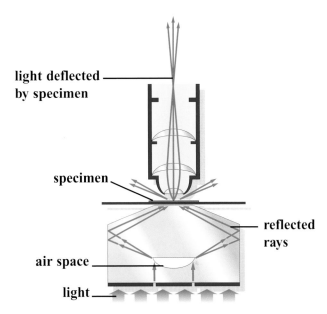

▲ Dark-field microscopes shine light from the side so the specimen is lit against a dark background.

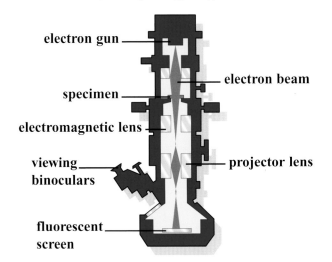

▲ With a TEM, an electron beam passes through the specimen onto a screen, which produces the image.

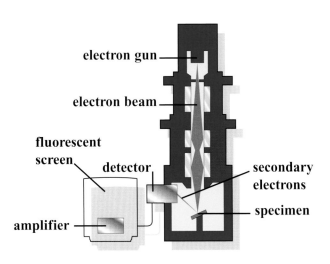

▲ With an SEM, an electron beam scans over the specimen, which sends more electrons to the screen.

lenses ever known, and the sharpness and enlargement of the images his instruments achieved has not yet been improved. Indeed, it needed completely new technologies to increase magnification still further.

DIFFERENT MICROSCOPES

Much less than a century ago, microscopes magnified only by using round glass lenses. Glass lens microscopes, called optical microscopes, are still in widespread use. However, there are three more kinds of microscopes that use different systems to achieve much greater magnifications: electron microscopes, scanning probes, and acoustic microscopes.

Optical microscopes

Janssen's original compound microscope had just two lenses—one object lens and one eyepiece. Modern optical microscopes have a single eyepiece, but they often have two or three object lenses.

▼ *Tiny mites are flash-frozen on their host with liquid nitrogen before being viewed in fine detail on a low-temperature scanning electron microscope (SEM).*

Typically, the first lens would magnify the object by four times (4 ×), the second raises magnification to 10 ×, and the third to 40 ×. If the eyepiece magnifies 10 ×, these lenses combine to give a magnification of 400 ×. Some of the most advanced research microscopes combine a 100 × object lens with 20 × eyepiece to give a magnification of 2,000 ×. This magnification can resolve tiny cells in the body.

Electron microscopes

Electron microscopes do not use light and lenses to magnify objects. Instead, they fire a stream of electrons at the specimen. The electrons bounce off the object or shoot through it. They then hit a fluorescent screen, creating an image of the object. An electron microscope can focus on something as small as 1 nanometer (½₅ millionth of an inch, or 1 millionth of a millimeter) and magnify it up to five million times.

There are two different kinds of electron microscopes. Scanning electron microscopes (SEMs) scan a narrow beam of electrons across the surface of objects and can magnify them up to 100,000 times. They can show things such as the internal

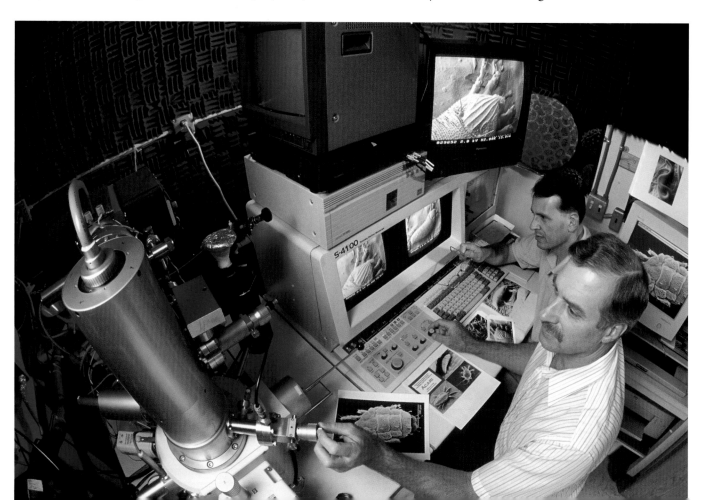

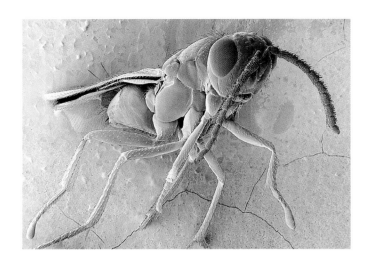

▲ *An SEM reveals the body of a species of wasp,* Tachinaephagus zealandicus, *in amazing detail. The wasp was introduced from Brazil to the United States to help control pest houseflies.*

structure of a bacterium. Transmission electron microscopes (TEMs) shine a broad beam of electrons through thin slices of an object to magnify things millions of times. They can reveal microscopic details such as the tiny individual structures in skin cells.

Scanning probe microscopes

The scanning probe microscope was developed by German-born U.S. physicist Erwin Mueller (1911–1977) in the 1950s. These microscopes have tiny needle-like probes that scan across the specimen. The idea is to provide a 3D image of atoms and molecules on the surface of the specimen.

One kind of probe microscope, called the scanning tunneling microscope (STM) was invented by International Business Machines (IBM) in Zurich, Switzerland, in 1981. The word *tunneling* refers to a phenomenon of quantum physics. An electrical current flows between the tip of the probe and the specimen. As the probe scans over the specimen, the current varies as it passes over each atom. From these variations, a computer builds a three-dimensional map of each atom.

In another kind of probe microscope, the atomic force microscope (AFM), the probe touches the surface of the specimen. As the probe scans, it follows the surface in three dimensions. These

movements are fed back to a computer to create the image. AFMs are being used to nudge into place tiny tubes of carbon atoms, called nanotubes, to construct the smallest-ever electronic chips.

Acoustic microscopes

Scanning acoustic microscopes use sound waves to construct images of microscopic objects. They work by sending out beams of sound that reflect off the surface of the specimen. The reflected waves are picked up by a receiver and then converted into an image. Scanning acoustic microscopes give relatively small magnifications, but they can see tiny opaque objects, without harming them, just as ultrasound machines see inside human bodies.

DID YOU KNOW?

In 2004, the United States Geological Survey (USGS) was awarded a patent for its underwater microscope system. This system will allow scientists to take magnified still images of, and film in minute detail, parts of rivers or seabeds.

Specimens and slides

Samples must be specially prepared to be viewed under microscopes. The preparation of specimens is such a vital part of the process that there is a whole branch of science devoted to it, called microtomy. Typically, specimens to be viewed under an optical microscope are sliced to a predetermined thinness to allow light to pass through them. They are then deposited on a thin strip of glass, called a slide. The finished slide is then placed on the microscope stage, which has a small hole, called an aperture, through which the specimen can be illuminated. Specimens for electron microscopy are more usually deposited on a thin copper grid and may be coated with gold if to be viewed in scanning mode.

See also: ELECTRON MICROSCOPE • RADAR

Microwave radiation

Microwaves are radio waves that form part of the electromagnetic spectrum. Microwaves are used in all kinds of electronic equipment, from radar and communications systems to microwave ovens, where their energy is used to cook food rapidly.

Microwaves are high-frequency electromagnetic waves. The difference between microwaves and other electromagnetic waves is in their wavelength—the length of one complete wave. Microwaves are very short waves. Their length ranges from $\frac{1}{400}$ inch (1 millimeter) to 12 inches (30.5 centimeters).

Microwave frequencies

The frequency of waves refers to the number that pass a particular point each second. The frequency of microwaves is usually measured in units called gigahertz (GHz). One gigahertz is a frequency of one thousand million waves per second. Microwaves range in frequency from 1 to 300 GHz. They form just a small part of a much wider range of frequencies called the electromagnetic spectrum. At the high-frequency end of this spectrum are cosmic rays, gamma rays, and X-rays. Then comes ultraviolet, visible light, and infrared radiation.

The bottom of the electromagnetic spectrum consists of radio waves. These are divided into eight ranges (bands). The top two bands are the EHF (extremely high frequency) and SHF (super high frequency) bands. Both the EHF and SHF bands fall within the range of frequencies known as microwaves. The next radio frequency band is the UHF (ultra high frequency) band. The top part of this band falls within the microwave range, too. Below this are the VHF (very high frequency), HF (high frequency), MF (medium frequency), LF (low frequency), and VLF (very low frequency) bands.

The different frequencies of the various kinds of waves in the electromagnetic spectrum give them their individual properties. Even different frequencies of radio waves can behave in quite different ways. The special properties of microwaves make them suitable for certain kinds of radio and television transmissions.

Microwaves in communications

High above Earth is a region of charged gases called the ionosphere. Longer radio waves are reflected by the ionosphere. A radio transmission can travel right around the world by bouncing between Earth and the ionosphere. Microwaves, however, pass right through the ionosphere. For this reason,

▼ *A communications tower with microwave transmitters and receivers. Microwave communications are transmitted around Earth using satellites.*

microwaves can be used for communication with spacecraft and for radar astronomy. It also means that, on Earth, microwave transmissions have a limited range. Beyond the horizon, microwaves may be bent or scattered by the atmosphere, so they are used mainly for "line-of-sight" transmissions.

The limited range of microwaves is used to advantage in television transmission. Television stations in different countries can broadcast on the same frequencies without interfering with each other.

Microwaves and radar

Microwaves are used in some radar systems. These work by sending out bursts of microwaves from an antenna. They travel outward until they strike an object that reflects microwaves, for example, an airplane. Some of the energy in the waves is then reflected back to the antenna. The reflected waves are used to form an image on the screen of a cathode-ray tube.

In this way, the microwaves reflected from an airplane can be used to form a bright spot on the screen of a ground radar system. The display on the screen represents a map, with the bright spot representing the airplane. The center of the screen represents the position of the radar station. Scales marked on the screen make it easy to see the range (distance) and bearing (direction) of the airplane from the radar station.

Range and direction

Radar equipment calculates the distance of an object from the time delay between sending out a burst of microwaves and receiving the reflected signals back. The greater the distance between the object and the radar station, the longer the time delay. Similar to other kinds of radio waves, microwaves travel at the speed of light—about 186,000 miles (300,000 kilometers) per second. On the screen, the distance between the airplane and radar station is translated using the time delay and knowing the speed at which microwaves travel.

Radar systems use microwaves because these extremely short waves are easily formed into a narrow beam. Longer radio waves tend to spread

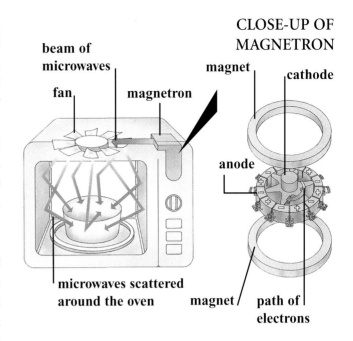

CLOSE-UP OF MAGNETRON

▲ *A microwave oven cooks food using microwave energy. An electronic device called a magnetron produces the microwaves. These microwaves are then bounced around the oven using a fan.*

out more. The waves must be concentrated into a narrow beam in order to find the direction of any reflecting object accurately.

For the same reason, space scientists sometimes use radar to study the Moon, meteors, and the planets. For this kind of long-distance radar, the microwaves are transmitted and received using huge, dish-shaped antennas called radio telescopes.

Microwave ovens

Perhaps the best known use of microwaves is to cook food. Microwaves are absorbed by water, fats, sugars, and certain other molecules which vibrate and produce heat. The waves heat the inside of the food, and not the surrounding air, greatly reducing the cooking time. Most types of glass, and some plastics, do not become hot because they do not absorb the microwaves. However, metals block out the microwaves and can increase the risk of radiation and arcing (electrical discharge).

See *also*: ANTENNA • BROADCASTING • RADAR • RADIO • TELEVISION

Mine, explosive

A mine is a hidden, exploding weapon used by the military to prevent the advance of enemy forces over the sea or across land. Mines are easy to make, cheap, and deadly.

Mining can be traced back to Biblical times. Fires and later explosives were set under fortified city walls by attackers who dug under the foundations like miners. The term *mine,* therefore, came to mean buried, or hidden, military explosives. In the 1860s, some mining was done in the American Civil War (1861–1865). But mines as they are thought of today—hidden explosive devices triggered by the enemy—were first produced and used in large quantities in World War I (1914–1919). There are mines for use on land and also naval mines used in water.

Land mines

Land mines are used to hold back advancing military forces and are usually detonated (exploded) when they are walked on or driven over. The first land mines were used during World War I, and were simply artillery shells buried just below the surface of the ground with their fuses pointing

▲ A U.S. Navy diver attaches a detonating charge to an underwater mine that has been discovered. Once the divers and any ships are clear, the mine will be detonated with a controlled explosion.

◄ U.S. marines from the 2nd Combat Engineer Battalion explode a line of land mines during a training exercise.

upward. When a tank rolled over the mine, the compressed fuse triggered the detonator and the shell exploded. Mines have since become more sophisticated. Modern land mines can be rigged to detonate in different ways, and they may contain many different explosive charges depending on their purpose. Chemical mines have even been developed, which release various poisonous gases into the air.

As well as being more sophisticated than old land mines, modern land mines can also be laid much more quickly. Early mines were laid by hand, which was time consuming. Modern minefields (the areas in which mines are laid) are laid, or "sown", by air or by specially adapted vehicles.

In general, land mines are either anti-personnel (AP) or anti-tank (AT). AP mines are designed to kill or injure soldiers. AT mines are designed to disable or destroy tanks and other vehicles.

Land mines are either triggered by pressure or a trip wire. Typically they are found either hidden on or just below the surface of the ground. Although AP and AT mines are basically the same, there are a couple of significant differences between them. AP mines are quite small—usually not much bigger than a can of soda. AT mines are larger (about the size of a large saucepan) and contain several times more explosives than AP mines. AT mines need enough explosive power to disable or destroy tanks and other military vehicles.

The second major difference between AP and AT mines is that AT mines require more pressure to detonate than AP mines. AP mines need only a pressure of between 10 and 35 pounds (5 and 15

▶ *Members of a United Nations bomb disposal team clear land mines in Africa. Land mines are often discovered by innocent civilians long after conflicts have ended. In many parts of the world, these mines cause thousands of injuries and deaths a year.*

kilograms) to detonate, whereas an AT mine typically requires between 350 and 750 pounds (160 and 340 kilograms) of pressure.

Anti-personnel mines

There are three basic categories of AP mines—blast, bounding, and fragmentation mines. Blast mines are the most common and the simplest type of mine. They are buried just under the ground surface and are triggered by someone stepping on a pressure plate on top of the mine. These mines are designed to destroy a person's foot or leg. They contain only a small amount of medium explosive, often Tetryl.

Bounding mines are more complex two-stage devices that bounce up to chest height before exploding. Bounding mines were first developed during World War II (1939–1945). The best known were the German "S" mine and the U.S. "Bouncing Betty." Modern bounding mines are still commonly called "Bouncing Bettys."

Bounding mines are buried with only small spikes, called ignitors, protruding from the ground. These mines are activated by trip wires or when someone steps on the mine. When a bounding mine is triggered, a propelling charge lifts the mine about 3 feet (1 meter) in the air. Only then does the main charge explode, letting loose a deadly shower of metal fragments. Bounding mines are designed to kill whoever activates the mine, as well as anyone in the vicinity of the explosion.

When fragmentation mines explode, they release deadly metal fragments in all directions, or in one specific direction (directional fragmentation mines). Fragmentation mines are effective within a radius of 650 feet (200 meters). Fragmentation mines can be simple blast mines or bounding mines. Bounding mines and fragmentation mines contain more powerful explosives than many AP blast mines, usually trinitrotoluene (TNT).

Anti-tank mines

AT mines are all blast-type mines. Since their aim is to destroy or disable tanks and vehicles, there is no requirement for bounding or fragmentation. AT mines contain powerful explosives, such as TNT or Composition B. AT mines are usually found on roads, bridges, and large clearways where enemy vehicles might travel.

Naval mines

Naval mines date back to the sixteenth century, when the Dutch filled a small boat with explosives and a clockwork detonation device and let it sail out to meet the Spanish Armada. Conventional "moored" mines first appeared in the nineteenth century. They are rounded metal containers filled with explosives and moored to the seabed in the same way as buoys. When a ship strikes one of the mine's "contact horns," a detonator explodes the main charge. This mine is called a contact mine.

Other naval mines include magnetic mines, which detect the magnetic field of a ship passing above. Pressure mines are placed in shallow water and detect the drop in water pressure as a ship passes above. Acoustic mines have microphones that listen for propeller noise. Modern naval mines also include the U.S. Captor mine, which is really an elaborate anti-submarine torpedo. The Captor's hydrophone (an underwater microphone) detects the sound of an enemy submarine (it can distinguish between friend and enemy). A computer control unit in the device then releases the torpedo and aims it at the intruder.

Mine clearing

Since mines are highly dangerous both on land and at sea, there are as many ways of finding and neutralizing mines as there are varieties of mine.

On land, the most effective way to detect mines is still the simple, manual way. Army engineers with long probes feel the ground in front of them as they advance. Many AT mines are "booby-trapped" against attempts to move them, so they are usually exploded once found.

Manual clearing is a very slow way to clear a large minefield, so many other methods have been developed. Magnetic metal detectors can be used against older mines, but nonmetal mines, often made of plastic, have become common since World War II. Tanks have been fitted with anti-mine devices that disturb the earth in front of them and explode the mine before they move onto it. More recently, plowshares have been used to push mines out of the way as a tank advances.

The constant battle of wits between minelayers and minelifters also takes place at sea. Moored mines are cleared by a boat called a minesweeper. It tows a cable through the water. A cutting device called a paravane is attached to the cable. It severs the mine from the mooring line so that it floats to the surface. There it can safely be detonated from a distance by gunfire. Magnetic mines are cleared by minesweepers in pairs, towing cables that carry an electrical current. The magnetic field set up in the water between the cables sets off the mine. Another way of avoiding magnetic mines is to neutralize a ship's magnetic field by passing energized metal cable around it while it is still in dock—a process called "degaussing." Acoustic mines are evaded with a special noisebox that can be towed in the wake of a minesweeper.

An unusual method of mine detecting that has been developed recently uses trained dolphins. Dolphins can carry fin-mounted sensors that record their response to underwater targets, enabling their handlers to pinpoint the location of suspected mines.

◀ *A U.S. Marine uses a dolphin to locate underwater mines in the Arabian Gulf.*

See also: BOMB • EXPLOSIVE • MISSILE AND TORPEDO • SONAR

Mineral

Geologists have discovered around 3,000 minerals in Earth's crust. Some, such as feldspar and quartz, which make up much of the rock called granite, are common. Others, such as diamond, emerald, ruby, and sapphire, are rare and extremely valuable.

All Earth's rocks consist of minerals. Only a few rocks are made from just a single mineral. Most are made from half a dozen or more. Minerals are naturally occurring chemicals that nearly always occur as crystals. Definitions of the word *mineral* vary, but most geologists agree that they are naturally occurring inorganic (nonliving) substances with specific chemical compositions and crystal structures. Coal is not always considered to be a mineral because it is made from the fossil remains of plants and is therefore organic.

Most minerals are compounds of two or more chemical elements. The mineral halite or rock salt is a compound of sodium and chlorine (NaCl). Just a few minerals, such as gold and copper, occur as free elements in nature and are called native elements.

Mineral formation

There are several thousand known minerals, but only about 30 are widespread. Most are present in rocks only in minute traces and are only easy to see when they become concentrated by geological processes. These concentrations provide the ores from which many metals are extracted and also the gemstones that are valued for decoration.

Mineral crystals form when liquids or molten solids in Earth's crust cool, so that chemicals within the liquids or molten

▶ *Quartz is the most common mineral in the world and is found in all but a few rocks, from granite to sandstone. Only rarely, however, does it form big, beautiful crystals such as these.*

solids solidify. The crystals grow as more atoms attach themselves to the initial structure—just as icicles grow as water freezes on them.

Some form when hot, molten rock from Earth's interior cools slowly. Some form from chemicals dissolved in watery liquids in the ground. Others form as existing minerals are altered chemically. Still others form as existing minerals are squeezed or heated by geological processes. Most of the time, the crystals form quickly and never grow very big— no bigger than a grain of sugar. Occasionally, they form very slowly, when single crystals may grow as big as telegraph poles.

Identifying minerals

Since there are so many minerals, geologists have figured out several methods of identifying them. Some minerals, such as magnetite, have magnetic properties, while others are radioactive. However, the most common ways of recognizing minerals are through their crystal structures, hardness and specific gravity, cleavage and fracture, color, streak and transparency, reflective and refractive properties, and their chemical composition.

CRYSTAL SYSTEMS

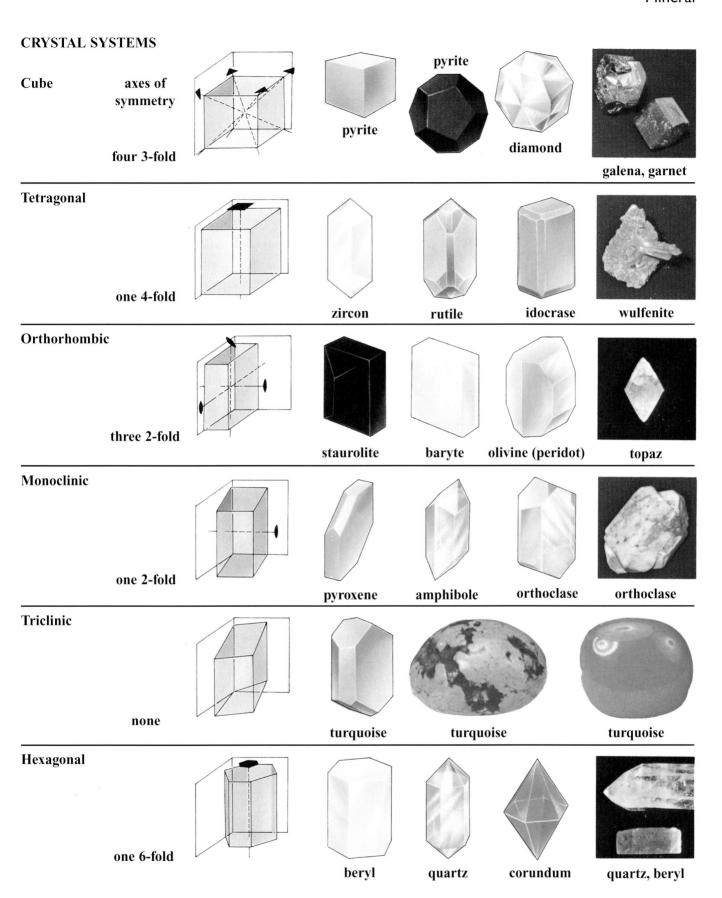

Cube — axes of symmetry — four 3-fold

pyrite

pyrite

diamond

galena, garnet

Tetragonal — one 4-fold

zircon   rutile   idocrase   wulfenite

Orthorhombic — three 2-fold

staurolite   baryte   olivine (peridot)   topaz

Monoclinic — one 2-fold

pyroxene   amphibole   orthoclase   orthoclase

Triclinic — none

turquoise   turquoise   turquoise

Hexagonal — one 6-fold

beryl   quartz   corundum   quartz, beryl

▲ *Mineral crystals are classified into six groups, called "systems," according to the symmetry of their shapes. Only rarely are perfect examples found in nature, but it remains a valuable way of identifying crystals.*

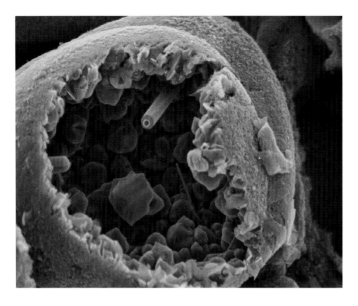

▲ *Graphite crystals formed in bubbles in industrial carbon look remarkably similar to crystals grown in geodes—bubbles in natural magma.*

Crystals

Crystals are a very particular kind of solid with regular, geometric shapes, smooth faces, and sharp edges and corners. Many crystals, though not all, are shiny or clear. Scientific analysis shows that they are made from atoms linked together to form a regular structure or crystal lattice.

The crystals in rocks often look chunky and shiny, but they rarely have the perfect shapes shown in drawings. All the same, scientists have discovered that all mineral crystals are formed in certain ways in a limited range of basic shapes. Individual crystals, for example, are all basically symmetrical. That means that they look the same shape from various different angles.

Although the symmetry is far from perfect in nature, almost all mineral crystals have what is called an axis of symmetry, which is an imaginary line through the center of the crystal. If a crystal is rotated around its axis of symmetry, it always appears the same from every angle. Crystals can be classified into six "systems" according to their axes of symmetry. The six systems are named after geometrical shapes: monoclinic, triclinic, cubic, tetragonal, orthorhombic, and hexagonal (or trigonal). Illustrations of these systems are shown on page 1009, with examples.

Some minerals, such as chalcedony, are called cryptocrystalline because their crystal structure is visible only through a powerful microscope. On the other hand, opal is amorphous (without regular shape). It contains no regular arrangement of atoms, so it is noncrystalline.

Usually, mineral crystals do not form separately. Instead, they grow together in a mass. When they form at the same time as the rock in which they are found, each crystal of the same kind tends to grow together in a particular way. So each kind forms a mass with a characteristic shape or "habit." Some grow in a branching, treelike way, called a dendritic habit. Others form needles. This growth pattern is called an acicular habit. Each kind of crystal tends to form in particular conditions, so each has its own typical habit. This helps identify the mineral.

Hardness, weight, and breaking

Some minerals, such as talc, are so soft that they crumble between your fingers. Diamond, the hardest natural substance, is so hard it will scratch glass. Hardness varies so much that it is another useful way to identify minerals.

The Mohs' scale, devised in 1812 by German mineralogist Friedrich Mohs (1773–1839), rates a mineral's hardness against common minerals of increasing hardness one to ten: talc (1), gypsum (2), calcite (3), fluorite (4), apatite (5), orthoclase (6), quartz (7), topaz (8), corundum (9), and diamond (10). Each mineral can scratch any other mineral with the same or a lower rating, but it cannot scratch minerals higher on the scale. Sometimes, substitutes are used to test minerals, such as a fingernail, which has a hardness of 2.5 on the scale. So a fingernail will scratch talc and gypsum but not calcite.

Some mineral lumps feel heavy compared with lumps of other minerals of the same size, because the specific gravity of minerals differs. Specific gravity is the ratio between the weight of a mineral and an equal volume of water. Gold has a specific gravity of 19.3. It is 19.3 times as heavy as the same volume of water. Quartz has a specific gravity of 2.65—the average for most minerals.

Some minerals can be recognized because they split, or "cleave," more easily in certain planes (directions) than others. There are four easily identifiable cleavage patterns: sheetlike or basal (mica), rodlike (orthoclase and feldspar), blocklike (halite), and diamondlike (fluorite).

Even when a mineral does not have clear cleavage planes, it may break, or "fracture," in a distinctive way, such as the conchoidal (shell-like), hackly (splintered), and jagged fracture patterns.

Color and streak

Color gives an instant clue to a mineral's chemical composition. The brilliant blue of azurite and the vivid green of malachite are unmistakable. The color usually comes from the main chemical constituents, and they are said to be idiochromatic. Peridot, for example, is green because of its high iron content. However, impurities—traces of other chemicals—can turn minerals all kinds of different colors. Minerals colored by impurities are said to be allochromatic. A few minerals, such as tourmaline, are pleochroic, which means they change color according to the angle at which they are observed.

No matter what impurities it contains, however, a mineral always powders to the same color. So if rubbed on a white porcelain tile, it always leaves the same colored mark or "streak." Hematite and magnetite are both black minerals. But magnetite has a black streak, while hematite has a red streak.

Light effects

Some crystals are as clear and transparent as glass, for example, pure calcite. Others are translucent or milky, which means you cannot see through them, but light shines through. An example is fluorite. Opaque minerals, such as galena, block out all light.

Minerals reflect light in different ways. Their surface gloss, or "luster," can be vitreous (shiny like glass), pearly, waxy, greasy, silky, or adamantine (sparkling like diamonds). Some minerals, such as opal, reflect shimmering rainbow colors. This effect is called iridescence. When light enters a mineral, it is refracted (bent). One mineral called Iceland spar (a kind of calcite) splits light rays completely in two. This effect is known as double refraction. If a sample of Iceland spar were placed on this page, two images of every word would be seen through the crystal. In some minerals, especially diamond, light rays are dispersed; that is, they are split into a band of the seven colors of the rainbow. In diamonds, this effect is known as fire.

Mineral groups

Minerals can be split into groups by their chemical composition. Silicates are by far the most common group of minerals. There are more silicates than all other minerals put together. Some, such as quartz and feldspar, make up the bulk of many rocks. They are basically metals combined with silicon and oxygen, the most common elements in Earth's crust, making up 75 percent of the mineral by weight. Most silicate minerals form when silicate-rich molten magma from Earth's interior bubbles up to the surface and solidifies. Besides quartz and feldspar, silicates include minerals such as amphiboles, micas, and pyroxenes.

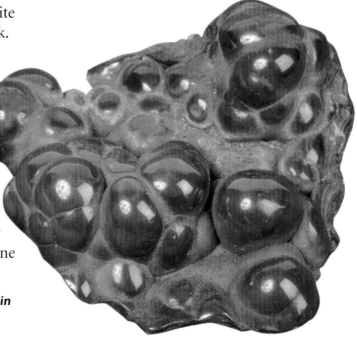

▶ *Malachite gets its distinctive green color from the natural reaction of copper when exposed to oxygen in the air. Malachite is perhaps the oldest known green pigment. It is sensitive to acids and to heat.*

Sulfides are generally brittle, heavy compounds of sulfur most often combined with a metal. They form from superheated water in veins (cracks) in the ground and include important metal ores, such as chalcopyrite (a copper ore) and cinnabar (a mercury ore). Sulfates are metals joined with a sulfate group, a partnership of sulfur and oxygen (SO_4^{2-}). Sulfate minerals include barite and gypsum. Another major group of minerals are the oxides, which are a combination of a metal with oxygen. Oxides include everything from dull ores such as bauxite (aluminium oxide; Al_2O_3) to rare gems like rubies and sapphires. Hard, "primary" oxides form deep in Earth's crust in hot magma and veins. Softer, "secondary" oxides form near the surface as sulfides break down and are a major element in soils. Carbonates are minerals that form when metals and semimetals join with a carbonate, which is a combination of carbon and oxygen (CO_3^{2-}). Most are formed by the alteration of other minerals on the surface. They are common in rocks such as chalk and limestone.

Minerals in rocks

Igneous rocks are formed from magma—intensely hot molten material from Earth's interior that sometimes emerges through volcanoes. Minerals

form as the magma cools and solidifies. When magma reaches the surface and cools quickly in the air, there is no time for crystals of the minerals to grow. So the minerals in surface-forming rocks, such as basalt, are very fine-grained. Deep down, though, the magma cools much more slowly, so crystals can grow much bigger. The minerals in deep-forming rock such as granite—quartz, mica, and feldspar—tend to be much coarser grained.

After the formation of granite, a mineral-rich liquid is often left over. This fluid enters open cracks in the rocks, where there is often space for large crystals to grow. Sometimes this fluid is forced into long cracks to form hydrothermal veins. These veins often contain rare minerals.

Sedimentary rocks are mostly formed from fragments of other rocks. For example, sandstone is a common sedimentary rock formed mainly from

◀ *This light micrograph of a slice of basalt rock reveals the grains of minerals from which it is made. The large, clear, grey crystals are quartz. The dark crystals are feldspar. All these are set in a fine-grained "matrix" known as phenocryst.*

▲ *Sometimes known as fool's gold, because it can be mistaken for gold, pyrite is a common sulfide mineral that often forms distinctive cubic crystals.*

sand. Sandstone is formed from fragments of granite, so the main mineral in sandstone is quartz, but feldspar and mica are also present. Sometimes sediments in riverbeds contain grains of rare minerals, such as gold, worn from veins in granite. Such deposits are called alluvial deposits. Some sedimentary rocks result from chemical action. For example, gypsum and rock salt are formed when the waters of seas and lakes evaporates.

Great heat and pressure below Earth's surface can metamorphose (change) existing rocks. The minerals in the rocks may be rearranged and recrystallized. In this way, some new minerals are formed that were not present in the original rock. The mineral garnet is formed by metamorphism. Weathered garnets, worn out of metamorphic rock, are also found in sand. Varieties of garnet are used as decoration for jewelry and other items.

Mineral resources
Minerals have played an extremely important part in human history—especially the metals copper, tin, and iron, which were used to make tools and weapons in the Copper, Bronze, and Iron Ages. These periods in Earth's history were important milestones in the rise of human civilization. Since then, many other metals have become important world commodities, including aluminum, gold, lead, platinum, silver, tin, and zinc. More recently,

radioactive elements used as fuels in atomic power stations, in particular uranium, have also become highly prized.

Most metals occur in combination with other elements in mineral ores. For example, iron rarely occurs as a native element but is found in such common ores as hematite, magnetite, and siderite. Similarly, most copper comes from mineral ores, including azurite, bornite, chalcopyrite, chalcosine, and cuprite (which is nearly 90 percent copper).

Economic minerals are those that contain enough metal to make mining and refining profitable. For iron ores to be profitable, they must contain between 30 and 60 percent iron. Silver is so rare that ores containing as little as one percent silver may be economical to mine. Geologists estimate that silver makes up only 0.00001 percent of Earth's crust and gold only 0.0000005 percent. By contrast, copper forms 0.007 percent.

Gemstones
Most minerals are dull colored, and their crystals are tiny. However, a few form richly colored, large, sparkling crystals called gemstones. Of the 3,000 or so minerals, only 130 form gems and even then usually only under certain conditions.

Diamonds are pure carbon, like coal, but they glitter like glass and are very hard because they have been transformed by enormous temperature and pressure. Most naturally occurring diamonds are very old indeed and were formed deep in Earth's core billions of years ago.

Some common minerals form gems in special conditions. For example, corundum is not a gemstone, but it can form rubies or sapphires. Ruby and sapphire are forms of aluminum oxide, just like bauxite. The red of ruby comes from chromium impurities, and the blue of sapphire from traces of titanium. Emeralds are a form of beryl turned green by traces of chromium and vanadium.

See also: CHEMISTRY • CRYSTAL • DIAMOND • ELEMENT, CHEMICAL • GEOPHYSICS • ROCK

Mining and quarrying

Mining and quarrying are ways of digging rocks and minerals out of the ground. Building stone, coal, diamonds, and metal ores are just some of the valuable materials that are taken from the ground in this way.

Mining involves the extraction of ores and minerals. The construction of some mines involve digging long shafts and tunnels deep underground, but mining takes place at all levels. Some mining is even done under the sea.

Quarrying involves digging large holes in the surface of the ground to extract stone, sand, and gravel in bulk, mainly for building material.

Mining and quarrying today are generally huge engineering projects, involving many millions of workers and massive equipment and machines for digging out the rocks and minerals.

The history of mining and quarrying

Mining dates back to the earliest days of human prehistory. The oldest mine in the world was sunk at Bomvu Ridge in Swaziland, Africa, more than 40,000 years ago. The mine was dug to extract the ocher used by the people of the time for body paint and funeral ceremonies. Then, throughout the New Stone Age (8000–2000 BCE), people were mining for flints to make arrowheads, knives, and sharp tools. In Britain and France, there are mine shafts up to 330 feet (100 meters) deep dug in soft chalk to get at the best flints. Amazingly, these deep shafts were dug with nothing more than bone and stone scrapers and the miners' bare hands.

When the values of metals were discovered some 6,000 years ago, mining became more important to people. Copper and tin were mined in many places and combined to make bronze. Experts believe the

▼ Miners in Bolivia are using a drill to extract tin ore. Tin mining is one of the oldest forms of mining, dating back some 5,000 years to the Bronze Age.

▶ *This gigantic hydraulic (fluid-driven) drill is being used to cut out coal. Coal seams are generally quite soft, so coal can usually be drilled rather than blasted out of the rock.*

ancient Egyptians were mining for copper on the Sinai Peninsula more than 5,700 years ago, and for iron almost 5,000 years ago. One method employed by the Egyptians was to dig tunnels through the hard rock using a technique called fire-setting. This meant heating the rock with a brushwood fire and then dousing the flames with cold water. The sudden cooling effect would crack the rock, and it could then be broken down more easily.

The Romans, too, did much to develop areas of mining. Using slave labor, they dug mines right across Europe and Africa. By the Middle Ages, mining for metals such as iron had become fairly well established, but methods remained very simple. This did not change until several centuries later. In 1556, German scholar Georg Bauer (1494–1555), better known by the name Agricola, wrote

> ## DID YOU KNOW?
>
> The Egyptians organized mining operations as early as 3000 BCE. They put thousands of prisoners to work, digging for precious minerals in deposits that lay near the surface. These minerals, including gold and silver, added greatly to Egypt's wealth.

the definitive work on mining—*De Re Metallica*. In his treatise, Agricola described in detail how to dig mine shafts and tunnels, how to break rock with a pick and hammer or by fire-setting, how to ventilate mines and pump out water, and how to bring out the minerals on trucks.

The coming of machines

The Industrial Revolution of the eighteenth and nineteenth centuries turned mining into a major worldwide industry. Industry demanded two things in particular: coal and iron. Coal was needed for iron-making and to provide the fuel that kept the new factories in production. Iron was needed to make machines and for construction.

Rapidly, the major coal deposits of Europe and North America became the focus of mining on a gigantic scale. Landscapes were transformed as mining companies gouged away at the surface to bring out poor-quality brown coal, or sunk shaft after shaft to reach deep deposits of higher-quality black coal.

The Industrial Revolution provided the demand as well as the means to dig deeper. There was machinery to circulate air, pump water, and hoist workers and materials. One key development was the use of gunpowder, and later dynamite, to blast rocks out. Since 1955, dynamite has been largely superseded by explosives made from a mix of ammonium nitrate (NH_4NO_3) and fuel oil and also by slurry explosives, which contain enough water to make the blasted rock flow out as a slurry.

Steam pumps and drills

Another milestone invention was the steam pump for pumping out water from mines. In 1698, English engineer Thomas Savery (c. 1650–1715) invented a pump to raise water using the suction created by condensing steam. Then, 14 years later, English blacksmith Thomas Newcomen (1663–1729) developed a much more efficient steam pump by incorporating a piston to provide the suction. In 1765, Scottish engineer James Watt (1736–1819) improved on Newcomen's design further by separating the piston from the condenser entirely. With pumps such as these, water could be pumped out quickly from huge depths, allowing deeper mines to be dug without flooding.

Another key step in the mechanization of mines was the development of mechanical drills to drill into the rock. In 1813, English engineer Richard Trevithick (1771–1833) invented a steam-driven drill. By the 1850s, steam-driven drills that moved the drill bit back and forth with a piston were widely used in mines. In 1853, hammer drills driven by compressed air, similar to the drills of today, were in use in Germany, and they soon replaced the steam-piston drills.

Mine safety

Deep coal mines remained extremely hazardous places to work, because of the frequent buildup of firedamp (methane gas; CH_4) in the mines. Firedamp is highly explosive and could easily be ignited by the miners' lamps which, at the time, were little more than candles. In 1815, English chemist Humphry Davy (1778–1829) and English inventor George Stephenson (1781–1848) both invented a safety lamp for miners to use. Davy's safety lamp surrounded the miners' lamp flame with wire gauze, preventing it from escaping and igniting any gas outside. This dramatically reduced

▶ *Open-pit mining has created some of the deepest humanmade holes in the world and employs some of the world's largest trucks. Although dwarfed by the hole itself, the trucks in the foreground of this photograph are gigantic.*

▼ *To extract metals such as iron from the ground, miners dig out ores, which are rocks containing metals in combination with other substances. Once the ore is mined, it must be refined to extract the metal.*

the number of explosions in mines, making it possible to mine much deeper. Even so, firedamp still remains a hazard in modern coal mines.

Although modern mining remains one of the most dangerous of jobs, it is much safer than it used to be. The machinery is so improved that jobs once done by people are increasingly being taken over by machines. This makes mines safer but puts people out of work. However, total automation, where machines do all the work, will be difficult to achieve in mining. The sites are large and the operations are varied and therefore difficult to control.

KINDS OF MINING

There are two main kinds of mining—surface mining and underground mining. Surface mining is by far the most common mining method and accounts for about 70 percent of all the world's mining production.

Surface mining

Surface mining is either "open pit" or "opencast." This is used when coal, ores, or stone lie close to the surface. The minerals are extracted by using explosives to blast away the "overburden," which is

▶ *This illustration shows a cross-section of a typical modern, highly mechanized coal mine.*

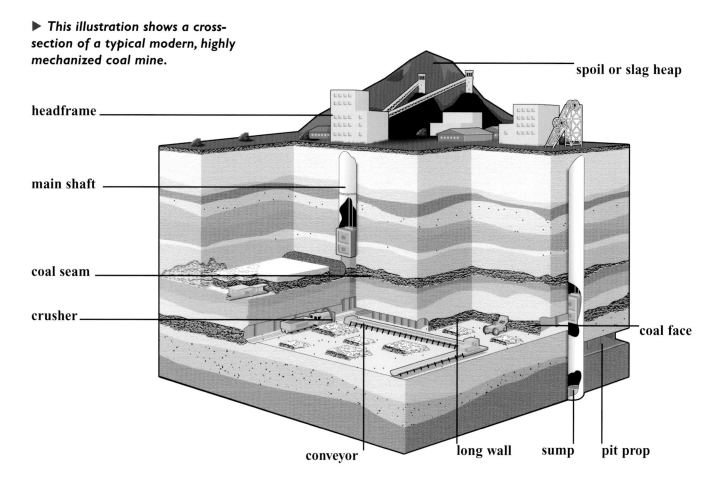

spoil or slag heap

headframe

main shaft

coal seam

crusher

coal face

conveyor

long wall   sump   pit prop

the layer of rock covering the deposit. The deposit is usually worked out in huge horizontal layers called benches.

Open-pit mining is a truly massive operation, moving more than 30 billion tons (27.2 billion tonnes) of material from the ground each year. The Kennecott copper mine in Bingham Canyon near Salt Lake City is more than ½ mile (0.8 kilometers) deep and 2½ miles (4 kilometers) across, making it the world's biggest humanmade hole. The trucks used to move all this material are huge vehicles, with wheels the size of houses.

Surface mining also includes borehole mining. Liquid is pumped down deep boreholes to dissolve the deposit. The solution is then pumped up to the surface, and the deposit is recovered. This method is used for mining salt and sulfur.

Another type of surface mining is marine mining. Special machines called dredgers with suction pump equipment regularly probe the seabed to a depth of 200 feet (60 meters) or more.

They bring up deposits ranging from diamonds to sand or gravel. Sand and gravel are mainly used as building materials and to rebuild shorelines.

Strip and placer mines

Another type of surface mining is strip mining. This is used to extract coal and minerals such as phosphate rock when they are found in flat layers close to the surface. Gigantic digging machines cut long strips or furrows to excavate the deposit, work the strip, then move on to another strip. Where the land is flat, the strips are deep ditches, called area mines. In hilly areas, the strips are notches running around the hillside, called contour mines.

By digging and moving on, strip mines and their spoils can quickly cover a vast area. Strip mining has a poor reputation, because it wrecks huge areas of the landscape. Nowadays, mining companies in the United States must demonstrate reclamation plans before they are allowed to excavate deposits from strip mines.

Not all minerals have to be excavated from solid rock. Heavy deposits of minerals such as gold, platinum, and silver are often found mixed with sand and gravel in what are known as "placer" deposits. Placer deposits have been deposited as rock and then broken down by weathering. These deposits can be more easily recovered.

On a small scale, gold can be recovered from placer deposits in streams by panning, which means washing out the sand and gravel in a pan, leaving the heavier gold grains behind. The famous California gold rush of 1849 began with the discovery of gold by panning.

Placer mining is not always small scale. In thick placer deposits, such as those on lake beds, the gravel and sand is hauled out by dredgers. Dredgers have endless chains of buckets that scoop out the material so that the minerals can be washed out in the same way as panning.

Quarrying

Bulky deposits lying close to the surface are excavated in large pits called quarries. Although there are many kinds of rock, there are just two kinds for quarrying purposes: crushed stone and dimension stone. Crushed stone is small pieces of rock, such as sand and gravel. (Sand itself is rock that has been crushed or broken down by the sea, by wind, or by chemical processes.) Dimension stone is any rock cut from the earth in large blocks, such as marble.

Crushed stone

Most quarrying is for crushed stone. Huge quantities of crushed stone are used in drainage systems, in road building, and under railroad tracks, as well as a raw material in the manufacture of concrete. Other uses of crushed stone include the manufacture of abrasives, crushed limestone in blast furnaces, and chalk as a whitening agent in paper manufacture.

▶ *The beautiful limestone used for many of the buildings on the island of Malta in the Mediterranean Sea must be cut into blocks, one by one, after the deposits are quarried from the ground.*

Dimension stone

In ancient times, almost all quarrying was done to obtain dimension stone. The ancient Egyptians were excellent at quarrying. To build the pyramids, they quarried huge blocks of limestone, often weighing up to 32,000 pounds (15,000 kilograms) and transported them over vast distances. The great pyramid of Khufu contains 2.3 million giant limestone blocks.

Today, many types of rock are quarried as dimension stone. Granite is used for buildings, memorials, and monuments. Marble is used for both building and sculpture. Slate is made into chalkboards and roofing tiles. Sandstone is used for building and for making grinding wheels.

Quarrying methods

With both crushed stone and dimension stone, quarrying begins in the same way. Once a suitably large deposit has been found, the topsoil and anything else covering the rock is cleared away. After this, however, quarrying methods differ for the two kinds of stone.

With crushed stone, vertical holes are drilled in rows using cable-tool drills. Explosive is packed into the holes, and the top part of each hole is packed with dirt or sand. Thousands of tons of broken rock may be blasted loose at one time. Machines then crush the rock down to size.

Extracting large blocks of dimension stone without breaking them needs tremendous care. The first step is to make vertical cuts in the rock to remove a large section of stone called a key block. The miners then separate individual blocks from the surrounding rock. For softer rock, a channeling machine is used. This machine consists of several steel bars with chisel edges clamped together. It cuts a channel several feet deep and about 2 inches (5 centimeters) wide. For harder rock, the most common method is to use a wire saw. If there is no handy natural joint at the base of the rock, it must be separated carefully from the rock bed by drilling horizontal holes and driving in wedges.

Underground mining

Underground mining is used when deposits lie far below the surface. Some high-quality coal is found deep underground. The gold mines of South Africa are the world's deepest mines, going down as far as 2 miles (3 kilometers).

The process begins by digging a main access tunnel. This may be a vertical shaft, plunging straight down through the ground to reach the deposit, or a horizontal or sloping tunnel, called an adit, which is dug into the side of a hill. Tunnels are then dug out from the main shaft toward the deposit. If the surrounding rock is strong, it will support the tunnels. If it is weak, the walls and roof of tunnels must be reinforced with steel pit props. Pitfalls are rare nowadays. Tracks are laid to carry material out in wagons, and elevators built to lift material (and miners) to the surface.

The deposits are worked in underground mines in various ways. In long-wall mining, for example, miners cut into the ore along a single long face. In room-and-pillar mining, they take out as much material as possible, leaving just narrow pillars of the ore to support the roof.

▼ *Blasting at mines is carefully controlled, maximizing the yield of valuable mineral deposits yet at the same time minimizing the impact on the environment.*

Digging and blasting

Coal is quite soft and can be chopped from the coal face by machine. First, long underground roadways are dug out mechanically, and then other machines are brought in to cut and load the coal.

Rocks containing metallic ores are much harder than coal, however, and the ores cannot be extracted using mechanical methods alone. After tunnels have been dug to expose the face, the ore has to be blasted from the rock using high explosives. Sometimes a method similar to the borehole technique is used to separate further the metal from the ore by dissolving it.

Similarly, gas can be taken from coal and piped to where it is needed, leaving the coal underground. This is done by digging shafts into the coal deposit, setting fire to the coal, and then collecting all the gas that is emitted.

Underwater mining

Deep under the world's oceans, there are areas of seabed covered with polymetallic nodules. These are small lumps of minerals, mainly manganese but also containing cobalt, copper, lead, nickel, zinc, and many others. They can be found in the Atlantic, Pacific, and Indian Oceans.

The minerals that make up the underwater nodules are found in traces in the seawater. They settle around a node on the seafloor, which can be a shell or a shark's tooth, and grow slowly into lumps that look something like potatoes. A nodule can be anything from a few inches (centimeters) to several feet (meters) across.

No one really knows exactly how much mineral wealth there is in this form. There are estimates of more than one trillion tons of nodules in the Pacific Ocean alone and that throughout the world the supplies of some minerals would not run out for centuries. The most valuable minerals found in this form are cobalt and nickel. The nodules usually contain about 30 percent manganese, but this is not very valuable because it is already found in many parts of the world. However, it might be useful for the United States, which has to import nearly all of its manganese.

▲ *Cutting out blocks of rock such as these dimension stones without breaking them is a surprisingly delicate operation. Each block is cut painstakingly from the ground using tough mechanical cutters and saws.*

Underwater mining involves dredging the ocean bed for small lumps of minerals. Many large mining companies have been working for years on developing economical ways of dredging these minerals from the ocean floor. Since the minerals exist at great depths, underwater mining is difficult. The process is likely to be costly, so large amounts must be brought up to make the operation worthwhile. One U.S. company has worked on ideas for mining at depths of between 1½ and 3 miles (3 and 5 kilometers). Ordinary dredgers are not usually designed to work at these depths.

Along with mechanical problems, there are political issues. The sea does not belong to any one country, but to all the nations of the world. However, only the wealthier countries could afford to mine the oceans. An international debate, known as the Law of the Sea Conference, has been going on for several years to try and overcome this problem.

See also: ABRASIVE • COAL MINING • GEOLOGY • OIL EXPLORATION AND REFINING • ROAD AND ROAD CONSTRUCTION • ROCK

Glossary

Alloys Mixtures of metals with other elements, often other metals or carbon, to make them stronger or give them certain qualities.

Anabolism Chemical reaction in which a complex substance is made from simpler ones. This leads to the storage of energy.

Atom The smallest particle of an element that has all the properties of that element.

Calculus The branch of mathematics that deals with rates of change in systems.

Catabolism All the enzymatic breakdown processes in an organism, such as digestion and respiration.

Corrosion Oxidation of metals or alloys on exposure to air, moisture, or acids.

Crystal Any solid object in which an orderly three-dimensional arrangement of the atoms, ions, or molecules is repeated throughout the entire volume of that object.

Element Any one of the fundamental chemical substances of which all matter is composed. Elements consist of atoms that contain the same number of protons and cannot be decomposed into smaller substances by normal chemical means.

Enzyme A protein that catalyzes a chemical reaction in a biological system.

Extrusion A molding process whereby a viscous molten substance is forced through a small hole.

Free fall Acceleration of a body under the sole influence of a gravitational field; that is, there is no air resistance or buoyancy.

Geometry The mathematics of the properties, measurement, and relationships of angles, lines, points, solids, and surfaces.

Gravity The natural force of attraction exerted by a massive body, such as Earth, upon objects at or near its surface, tending to draw the objects toward the center of the body.

Hormones Chemicals secreted into the blood by ductless glands and carried to specific cells, organs, or tissues to stimulate chemical activity.

Inertia The state of matter that prevents it from moving when it is at rest or from stopping or changing course when it is moving in a straight line, unless an external force is applied.

Kinetic theory The mathematical description of the behavior of gases.

Latitude Part of the grid system used to locate points on Earth's surface. Lines of latitude form parallel circles around Earth, concentric to the Poles.

Longitude Part of the grid system used to locate points on Earth's surface. Lines of longitude form semicircles joining the Poles.

Meteorology The study of Earth's atmosphere, in particular weather forecasting.

Minerals Naturally formed solids with specific crystal structures and definite chemical compositions.

Orbit The path described by one body, such as Earth, in its revolution around another, such as the Sun, as a result of their mutual gravitational attraction.

Ores Minerals from which metals can be extracted.

Proteins Large organic molecules containing nitrogen. Proteins are formed from combinations of amino acids.

Radiation Energy radiated or transmitted as rays, waves, or in the form of particles. Visible light and X-rays are examples of radiation.

Radioactivity The spontaneous disintegration of unstable nuclei, which is accompanied by the emission of particles or rays.

Refining Any one of a number of processes used to free metals from impurities or unwanted material.

Transistor An electronic semiconductor device, often used as a current amplifier.

Trigonometry The branch of mathematics that deals with the relationships between the sides and the angles of triangles.

Ultrasound Sound waves that have a frequency beyond the limits of human hearing.

X-ray Short-wave electromagnetic radiation produced when speeding electrons hit a solid target.

Index

Page numbers in **bold** refer to main articles; those in *italics* refer to illustrations.